The Second Law of Life

Energy, Technology, and the Future of Earth As We Know It

The Second Law of Life

Energy, Technology, and the Future of Earth As We Know It

John E. J. Schmitz

William Andrew
Publishing

Norwich, NY, U.S.A.

Copyright © 2007 by William Andrew, Inc. No part of this book may be reproduced or utilized in any form or by any means, electronic or mechanical, including photocopying, recording, or by any information storage and retrieval system, without permission in writing from the Publisher.
Cover by Hannus Design.

Library of Congress Cataloging-in-Publication Data

Schmitz, John E. J.
 The second law of life : energy, technology and the future of earth as we know it / John E.J. Schmitz.
   p. cm.
 Includes bibliographical references and index.
 ISBN-13: 978-0-8155-1537-1 (978-0-8155)
 ISBN-10: 0-8155-1537-5 (0-8155)
 1. Entropy. 2. Thermodynamics. I. Title.

QC318.E57S36 2006
536'.73--dc22
                                    2006020372

Printed in the United States of America
This book is printed on acid-free paper.
10 9 8 7 6 5 4 3 2 1

Published by:
William Andrew Publishing
13 Eaton Avenue
Norwich, NY 13815
1-800-932-7045
www.williamandrew.com

Sina Ebnesajjad, Editor in Chief (External Scientific Advisor)

NOTICE
To the best of our knowledge the information in this publication is accurate; however the Publisher does not assume any responsibility or liability for the accuracy or completeness of, or consequences arising from, such information. This book is intended for informational purposes only. Mention of trade names or commercial products does not constitute endorsement or recommendation for their use by the Publisher. Final determination of the suitability of any information or product for any use, and the manner of that use, is the sole responsibility of the user. Anyone intending to rely upon any recommendation of materials or procedures mentioned in this publication should be independently satisfied as to such suitability, and must meet all applicable safety and health standards.

Dedicated to Pieternel, Lucas, Juliette, Emmeline and Jasper

Contents

Contents vii

Foreword xiii

Preface xv

PART I The Birth of a Beautiful Theory: Thermodynamics 1

1. So What Is All This Talk About Entropy? 3

A few simple questions 3

A look at our travel itinerary 7

2. The Science of Heat and Work: Classical Thermodynamics 9

Historical context 9

How easy is it to go back? 15

Heat, energy, and mechanical work 20
    What do you mean when you say it's warm in here? 20

What is heat, what is work, and what is energy? 23

Entropy and the Second Law of Thermodynamics 28
The Carnot cycle 28
The moment supreme: entropy is invented 34
Entropy and reversible and irreversible processes 37
Entropy changes in spontaneous, isolated processes 41
The First and Second Laws combined 43

Perpetual motion and engines 43

Entropy and the direction of time 47

Let's take a break 48

3. Much More About Entropy 51

Do we really understand what entropy is all about? 51

History of the acceptance of the existence of the atoms in physics 52

Statistical Thermodynamics: macroscopic and microscopic views 54
It's all about probability 54
Connecting entropy with atoms and molecules 57
The Second Law when the systems become real 60

Entropy and the direction of time: reprise 64

Point of zero entropy and of zero absolute temperature 66

Boltzmann's struggle with the scientific community 68

Energy efficiency and some conclusions 69

4. Link of Thermodynamics to Modern Physics 71

Three men and thermodynamics 71

Why couldn't Newton's mechanics explain everything? 72
Dark clouds for classical physics 72
Black body radiation 74
The photoelectric effect 77
The Michelson-Morley Experiment 79
The connection between the classical mechanics of Newton, the Quantum Mechanical Theory, and the Special Theory of Relativity 80

Thermodynamics at the birth of modern physics 81
Boltzmann's heritage 81
What Planck thought about thermodynamics 82
What Einstein thought about thermodynamics 83
What Erwin Schrödinger thought about thermodynamics 86

The interpretation of time and its direction 88
What is time? 88
Einstein's interpretation of time 89

The influence of modern physics on thermodynamics: does relativity change entropy? 91

PART II Entropy and Our Society, Our Culture, Our Planet, and Our Universe 93

5. Entropy, the Economic Process, and the World's Environmental Problems 95

General environmental trends 96

How entropy plays a role in the economic process and re-defines concepts such as efficiency and sustainability 101
Relationship between thermodynamics and economic processes 101
Example of an economic process and the Entropy Law 103

A plea for a redefinition of efficiency and sustainability 106
Transformation of terrestrial resources from available to non-available 109
Summary of entropy and economy 110

6. Energy, Entropy, Life, and Heat Death 113

The contradiction between the thermodynamic push for chaos and the tremendous degree of molecular and biological organization 113
Chaos and life 114
The statistical nature of physical laws; or, to make something happen, atoms and molecules need to work together in large groups 116
Life and entropy 117

Entropy and the food chain. 121

Entropy and the planet 121
Energy and Entropy of the food chain 129

Heat Death 135

7. The Use of the Concept of Entropy in Other Sciences 137

Entropy and electrical communication 138
A brief history of electrical communication 138
Claude Shannon, the "inventor" of modern electronic communication network theory 140

Maxwell's demon 150

Use of the concept of entropy in other nonscientific fields 154
Entropy in the discussion of Christianity 154
The concept of Entropy and art 158

Epilogue 161

Appendix I. Two More Laws of Thermodynamics? 163

Appendix II. Another Way of Looking at Entropy 165

Appendix III. How Does the Gas Heat Up in the Air Pump? 167

Appendix IV. Will Reshuffling a Deck of Cards Change the Entropy? 171

Appendix V. How Much Does the Entropy Change in a Case of Gas Expansion and Gas Mixing? 173

Appendix VI. Thermodynamic Timeline 179

Appendix VII. Can the Human Body Be Considered a Heat Engine? 181

Contents   xi

Appendix VIII. Ways to Concentrate Energy: Nuclear Energy, Photovoltaic Cells, and Fuel Cells 183

Nuclear energy 183
Can photovoltaic cells provide the earth with a sustainable energy source? 186
How do solar cells work? 187
Fuel cells 191

Appendix IX. Qualitative Definitions and Descriptions of Entropy 195

Appendix X. Some Simple Calculations and Interesting Numbers 197

References 199

Index 205

Foreword

It is thought that of all the animals on planet earth, there is only one that can build a fire and has developed a realization of itself so that it can ask and answer the questions: Who am I? What is this fire that burns within me and before me? Why does the flame rise, and why am I warmed before this fire of time?

John Schmitz's book on thermodynamics is designed for the mature general science reader who has developed a general knowledge of the physical science literature that does not require mathematics beyond the arithmetic of writing a bank check. The overall objective of this short book is to introduce the reader to the thermodynamic concept of entropy and its many ramifications ranging from the micro-quantum world to the gross dynamic relativity construction of the universe. To prepare the reader for this entropy concept he lays down a foundation which closely follows the early historic development of thermodynamics. In preparation for reading this book, one should first carefully read through the table of contents. Dr. Schmitz makes statements and/or asks questions which he then answers in the text of the book, drawing the reader into his web of understanding which demonstrates the beauty and his love of thermodynamics. One very quickly realizes that in writing this book the author has given quality time in considering carefully the answers to his questions. There are footnotes that are well worth reading which amplify selected points including historic events with specific dates. You will find yourself going back to the table of contents and index pages to pick up action items in your reading. The reader should come to the realization that we do not have an energy crisis as energy can neither be created nor destroyed but only transformed.

Rather, we have an entropy crisis wherein we are using up well ordered fuel materials to generate heat, which is used to produce some product and/or to do a little useful mechanical work, and there is a residue of less ordered materials such as exhaust gasses. The examples of entropy are generally well written and are illustrative of the wide usage of the concept. The author's example from biology is noteworthy as it first appears that biological living systems seem to violate the laws of physics, but just as it requires energy to operate a refrigerator to make low entropy ice cubes from water (a high entropy state) it likewise requires an expenditure of energy to maintain the living state of matter.

This work should be required reading during the first weeks of a formal thermodynamics course while students are being introduced to the conjugate couples* of work and heat. The book provides the basic historic foundation of the main pioneers in classical thermodynamics concepts and their relationships to the modern physics formulations of Planck's quantum mechanics, and to Einstein's relativity ideas. Professional educators in thermodynamics will enjoy recommending this book to their non-science friends. Also, the reading of this book by the general public should generate a better educated public which should be able to maintain higher quality discussions using thermodynamics concepts as applied to political, social, religious, and economic problems of the future.

Gerald A. Kitzmann, Ph. D.
Professor & Chair Emeritus
Department of Physics
State University of New York

* Callen, Herbert B., *Thermodynamics and an Introduction to Thermostatistics*, 2nd edition, 18th printing (Wiley, New York, 1985).

Preface

From 1980 through 1984, I worked on my PhD thesis, which dealt with the rather arcane topic of the energetic properties of certain chemical compounds. The branch of science that describes those energetic properties in terms of heat and work is called thermodynamics, and it is how I became familiar with the concept called "entropy." Through the years I have noticed that the concept of entropy was used in many more fields besides physics and chemistry. Struck by the simplicity of thermodynamics, the fact that it had remained virtually unchanged for decades, and that the theory quietly determines many aspects of our daily lives, but not realized by many, I got the idea to write a book about it for the non-scientist.[1]

So what is the purpose of this book, and what will the reader be left with? Succinctly stated, this narrative describes the discovery of an extremely powerful (and in essence simple) theory that fundamentally influences our society and our everyday lives. In doing so, this volume sketches the historical, social, and economical contexts that surround the

---

[1] However, also those with a background in thermodynamics obtained during their engineering studies will find many topics they were not aware of such as in the social and economic fields outside chemistry and physics where the concept of entropy has been used.

development of thermodynamics. Without being subjected to a lot of formulas,[2] the reader will gain a good impression of how the scientific and scientific and philosophical communities dealt with the entropy concept after its introduction, and how insight on the subject gradually increased. Chapters 1 and 2 follow the historical development of the thermodynamic theory, visiting the Second Law and First Law of Thermodynamics in chronological order – an approach that I believe will give readers a good sense of the problems that scientists wrestled with more than a century ago.[3] (For instance, it was not until late in the 19th century that scientists began thinking of heat as a kind of fluid called "caloric," comparable to the fanciful "aether" which was thought to transmit light waves in a vacuum.) In addition, readers will learn that thermodynamics theory was crucial to the birth of quantum mechanics and relativity, two earth shocking theories that transformed classical physics into modern physics.

This book also will use familiar situations to illustrate theoretical concepts – for example, why can't I get better gas mileage out of my car? How are the 10,000 kilo joules (which is 2500 kilo calories)[4] adults consume daily used in the body? What is the difference between heat and energy? Why does an air pump heat up when I use it to inflate my bicycle tires? Of course, a lot of attention is paid to steam engines, those industrial wonders that first inspired the study of thermodynamics, and we'll also look at how thermodynamics reaches into areas outside science, such as art and religion.

The book is divided in two parts. In Part one, classical thermodynamics is discussed with a review and description of fundamental phenomena such as temperature and heat, followed by a detailed analysis of how mechanical theory (which deals with the laws of locomotion) developed over time, from Aristotle and the ancient Greeks to Newton, who completed the theory around 1600. Despite its tremendous success in explaining the orbits of planets and satellites (among other things), the mechanical theory failed to account for simple observations, such as the direction of heat flow and the occurrence of

---

[2] Although we cannot escape a few very fundamental ones (à la $E=mc^2$).
[3] "By following the historical development of the subject usually more knowledge can be gained than by inspecting the polished final product", quote from Walter Moore in his book Physical Chemistry.
[4] We will abbreviate joule by J, calories by cal, kilo joule by kJ and kilo calories by kcal

irreversibility (for instance, the forward-only direction of time). All the laws of Newtonian mechanics allow processes to go in either direction, and therefore cannot deal with these phenomena. We also touch several times on the direction of time and relate the past and future of the universe to the law of entropy, showing what the final consequences may be in the long term – such as the "Heat Death" envisioned by Helmholz. Of course, we cannot ignore the colorful history of perpetual engines, and will portray a few of them.

Then we extend our journey by considering the development of physics between 1800 and 1900, and we focus on a key interpretation of entropy developed by Ludwig Boltzmann around 1900. Boltzmann's statistical mechanical theory clarified the relationship between entropy and the degree of organization of the atoms and molecules of a given system. Also, we seek to understand whether the birth of the new physical theories around 1900 – the quantum mechanical theory of Max Planck and the relativistic theory of Albert Einstein – had any impact on the theory of thermodynamics and whether, in turn, thermodynamics played any role in the creation of these theories. Before getting the answers to these questions, the reader receives a quick course – in layman's terms – in the basic principles of both theories, and what led Planck and Einstein to create them.

In Part Two, we describe the impact of thermodynamics on our world. Entropy always increases; we can only minimize the amount of entropy production by making sure that the processes stay as close as possible to reversibility, which often means that processes must proceed at very low speed. Much of the energy wasted in modern industrial economies is the price we pay for speed. The high degree of irreversibility in many production processes (where high speed is often used in order to maximize "performance") is in fact accelerating the entropy production of our world. As a result, compared to the medieval era, we are more rapidly approaching the heat death situation which can also be taken as a measure of how quickly we are degrading our environment and essentially leaving less entropy for our children to work with. We also will show how environmental economists have used the concept of entropy and what conclusions they have drawn.

A lot of progress in the development of thermodynamic theory was made by physicians such as Clausius, Mayer, and Helmholtz, and not by physicists. That was partly because the doctors wanted to understand the reasons behind the production of animal heat. At some point, there was an interesting theory that the friction of the blood stream in the veins produced body heat! It took some years before it became clear that combustion was the actual source. Undoubtedly, the reader will wonder how, despite the natural push for chaos engendered by the Second Law of Thermodynamics, nature has nonetheless created tremendously well-organized forms of life (DNA, for example). Sometimes it seems that Mother Nature is tending toward the creation of extremely complex systems, in defiance of entropy. The answer to this paradox is that we must first properly define the boundaries of the system we are considering. In this case, the boundaries of that system could be the entire world. When we do this correctly, we will conclude that the creation of life comes at the cost of available energy elsewhere in our system. In other words, the creation of life extracts lots of energy from other parts of the system, and so increases entropy.

Finally, we will go to quite different fields where entropy has played a role. Since Boltzmann's invention of the statistical mechanical model and interpretation of entropy, many people from widely divergent disciplines have speculated on how entropy can be used and should be interpreted, even in such unlikely areas as communication theory, religion, and art. The Appendices at the back of the book contain further examples and illustrations, such as the entropic cases for solar and nuclear energy, and whether the human body can be considered a heat engine.

In summary, thermodynamics is a convincing illustration of the power of the human mind when applied to a very fundamental field, namely the understanding of energy transfers. As Einstein has said, it is safe to assume that this theory will live on forever, and will continue to provide our society with insights and guidance in energy conservation and many other areas.

Throughout the book I have highlighted opinions of experts while trying not to give my opinions or interpretations on the subject. The reason is that what can be found in the mainstream literature is clear enough and more than sufficient for what we want to achieve here: to explain the birth and use of the concept of entropy. No need for more interpretations. It is my humble hope that the reader will enjoy the book as much as I have enjoyed carrying out the research and writing it.

While preparing the manuscript I was encouraged and helped by a few people to whom I would like to express my sincere appreciation: Prof. Dr. Hilde Van Gelder (Catholic University of Leuven), Dr. Mart Graef (Philips Semiconductors), Drs. Martin Heerschap (General Electric Financial Services), Dan McGowan (SEMATECH) Dr. Paul Newman

(Univ. of Texas), Ron Schmitz, Lucas Schmitz and Prof. Dr. Ir. Jan Steggerda (emeritus professor Catholic University of Nijmegen). Also, very stimulating inputs and ideas were given by Millicent Treloar from William Andrew Publishing.

John E. J. Schmitz
Kasterlee, Belgium
July 2006

PART I

The Birth of a Beautiful Theory: Thermodynamics

In the beginning of all things there was the endless universe, that the ancient Greek writers called the Chaos. Chaos contained already the fundamental elements of which all things consist: earth, water, air, and fire.[5]

---

[5] From "Die Schönsten Sagen des Klassischen Altertums", Richard Carstensen, Ensslin & Laiblin Verlag, Reutlingen (my translation).

1
So What Is All This Talk About Entropy?

A few simple questions

Let's begin with a few, seemingly unrelated questions: Why does heat always flow from hot to cold? We see this phenomenon every day and consider it unremarkable, so we never wonder why it's so. But consider a cup of hot tea sitting on a table. We all know that the tea eventually will cool down to room temperature, but why doesn't the opposite ever happen? Why doesn't a cold cup of tea ever become hot spontaneously, without being actively heated? Now, imagine that we place the hot cup of tea in a box whose walls do not allow any heat or material to pass in or out (we're creating an isolated system, discussed more fully in Chapter 2), see Figure 1.1. If we wait long enough, the tea will assume the same temperature as the air in the box, and the temperature of the air in the box will go up slightly. However, the total amount of heat energy inside the box, including the cup of tea, will remain the same. Why does this happen?

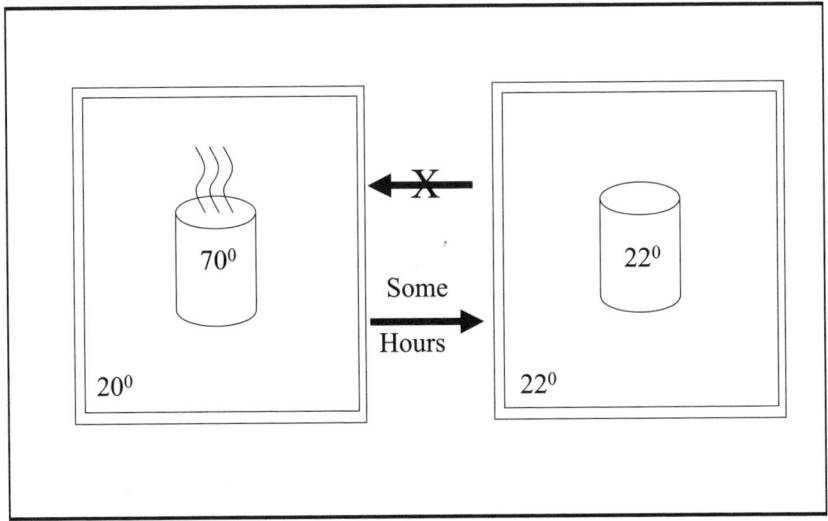

Figure 1.1 After a few hours the hot tea will have cooled down to room temperature. The opposite effect never happens spontaneously.

Next question: Why don't our cars get better gas mileage? Everyone knows that a lot of the energy (about 70%) contained in gasoline is wasted in the form of heat, rather than being converted into useful mechanical work that drives the wheels. That heat waste requires our cars to have extensive cooling systems, the most visible component of which is the radiator. The answer to the question of why the energy enclosed in gasoline cannot be converted entirely into mechanical energy is not obvious to us at this point.

Another question: Why does time, unlike many other circumstances of life, only go forward? Wouldn't it be fantastic if we could go backward in time and change history – our own, or the world's? (Unfortunately, as we'll find out, there's a fundamental reason why we can't.)

Essay question: Is nature chaotic, or orderly? Look at the world around you, and note how nature seems to possess two contradictory properties – the tendency to disintegrate, and the drive to organize. On

1: So What Is All This Talk About Entropy?

one hand, it seems that everything you put in order begins to degenerate into chaos as soon as you leave it on its own. For example, let's say you build a wall in your back yard from loose rocks or un-mortared bricks. Then you head off for five years to study archeology in Greece and Egypt. When you come home, chances are good that your carefully constructed wall resembles the archeological ruins you've been studying: your structure is gone and only scattered building materials remain. Or, imagine that your son holds a can full of marbles. Suddenly he drops the can, and all the marbles tumble onto the floor. Certainly you don't expect the marbles to somehow arrange themselves into a perfect pattern, but instead to go all over the room. So what's the fundamental reason behind this natural tendency towards chaos? Why don't the stones remain on top of one another indefinitely, and why don't marbles ever form a nice pattern when they drop on the floor? Why is there always this creation of a mess?

On the other hand, nature is also a builder. Millions of years of evolution have created living organisms that are the epitome of organization – consider the genetic code in DNA, for example. It's hard to understand how nature, which tends to be chaotic, can also exhibit such well-organized behaviors.

Bonus questions: What is energy? What is heat, and how does it relate to energy? Everybody uses these words almost daily. But can you explain the difference between heat and energy?

These questions seem simple at first, but answering them turns out to be tougher than we thought. Fortunately, we don't have to do it ourselves. Over the last 150 years, brilliant people making similar observations, and asking like-minded questions, have actually found the answers. These answers have come together to form the concept of entropy, as explained in the theory of thermodynamics, a word derived from the Greek words *thermos* ("heat") and *dynamos* ("force"). Thermodynamics studies the relationship among energy, heat, and work.

The birth and development of thermodynamics were driven by economics – namely, the need to better understand the efficiency of steam engines. After many centuries of using the laws of mechanics to study the movement of objects in the sky[6], scientists finally turned to work on more terrestrial pursuits, notably improving the performance of

---

[6] We mean here the mechanical theory of our solar system which started with Aristotle and was completed by Isaac Newton in 1687. See Chapter 2 for more detail.

steam engines. While the direction of heat flows had been well known to humanity for thousands of years, the underlying description of that phenomenon – that is, the theory of thermodynamics – began taking shape only 150 years ago[7]. As we'll see in Chapter 2, steam engines were among the most important drivers of the Industrial Revolution, and the new theory of thermodynamics helped explain how they really worked. That improved understanding led to better engines, which helped lower production costs and created cheaper goods, and so had a tremendous influence on history and on the quality of people's lives.

One of the most intriguing concepts emerging from the thermodynamic theory is entropy. As a relatively abstract concept, entropy has received considerable attention since its conception in 1865 by Rudolf J.E. Clausius, a German physicist; yet it usually has been associated with mystical and esoteric thinking and has sometimes been used and interpreted in questionable ways. The fact is, entropy has a major impact in many areas of our daily lives, from the relationship of heat and labor that leads to machines which can transform heat into mechanical work, to philosophical discussions on the creation of the earth and the universe and the passing of time.

When I ask my friends and neighbors what they know about entropy, most of them admit they don't know anything. Some vaguely remember something about chaos, but don't know exactly how chaos relates to entropy. That's why I wrote this book – to show you some of the truly breathtaking results of the theory of entropy – and to do it so you don't need to know much more than high school math to understand it all.

Surprisingly, in the vast amount of literature that has appeared on entropy, only a few words are devoted to the origin of the word itself. Clausius, who lived from 1822 to 1888, introduced the concept. He derived the word from two others: *energie* (German for "energy") and the Greek word *trope*, which means "turning" or "turn". Clausius discovered that, when used to generate mechanical work, a certain amount of energy was transmuted into a form that could no longer be used to produce work. He tried to capture this observation in the word *entropy*. Chapter 2 explores his findings in more detail, in context with the social, economic, and political environments in which Clausius and other scientists lived and worked.

---

[7] See for more historical details Georgescu-Roegen, 1971

1: So What Is All This Talk About Entropy?

Before going on, let me make a suggestion on how to read this book. To understand what's happening, you should read Chapters 2 and 3 first, in serial order. These two chapters give you the necessary background to understand all of the other chapters. Chapter 2 explains the fundamentals of "classical" thermodynamics and is perhaps the most difficult one, while Chapter 3 expands on the fundamentals of the concept of entropy, using the principles of atomic mechanics. This chapter is the heart of the book, and is pivotal to understanding what follows. Although the number of formulas is kept to an absolute minimum, we cannot totally avoid them – but don't be afraid. Close to each formula you will find a clear graphic to explain what it means. Take a few moments to study and understand the graphics, and you'll be fine. Chapter 4 can be skipped and read later. The remaining chapters can then be read in any order. Those who would like to see some more material and applications of the thermodynamic theory can refer to the appendices, but that is not necessary to understand the rest of the book. Think of the book as a history of ideas, a journey through time, as we explore an important concept – entropy – and its effect on us.

A look at our travel itinerary

In Figure 1.2, I have sketched the different "stations" we'll stop at during our journey of the mind. Between the brackets you will find the names of the leading scientists whose work we'll investigate, and the associated chapters.

We'll start our ride with a question in mind: How does entropy work; what is the magic here? Following the tracks to an answer will take us on a trip that begins around 1820 in France, then winds through time and other countries, where great minds devoted much of their lives to understanding the fundamentals of thermodynamics and entropy. But our journey is more than historic; it carries us into the present to a much better understanding of many things we observe in our everyday world. We'll also cross the boundaries of many fields of interest, learning that the concept of entropy also has influenced chemistry, relativistic physics, mathematics, environmental and energy conservation, information theory, cosmology, economy, and biology. And we'll find out how entropy remains at the heart of both scientific and non-scientific debate. Though not everyone agrees on the ways entropy has been interpreted in these diverse fields, or on the conclusions reached in its far-flung applications, the story of entropy is a fascinating study of the power of an idea to illuminate human experience.

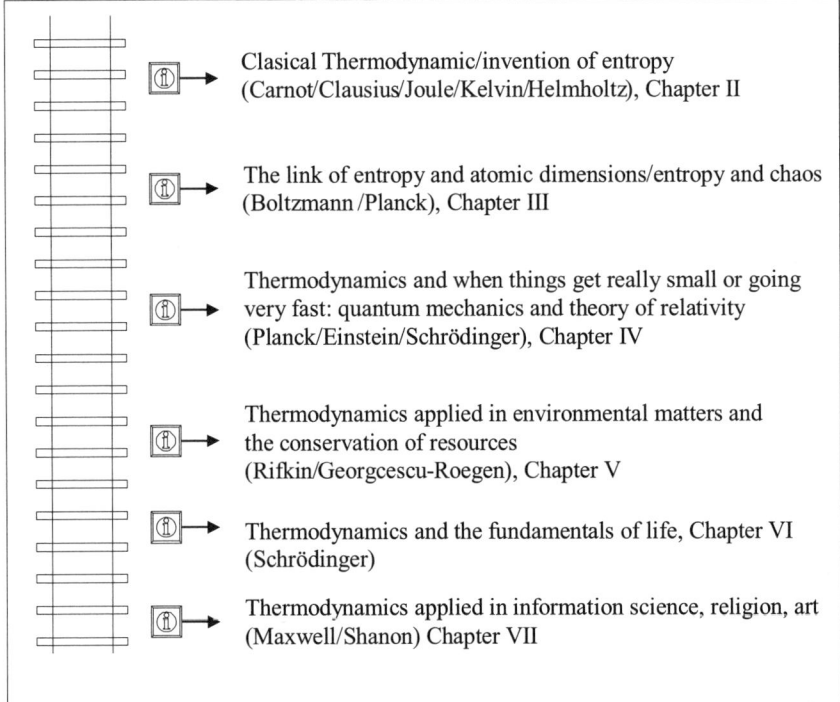

Figure 1.2 A travel itinerary for this book. In the first four stations the basics of thermodynamics and of entropy will be established. In Chapters 5, 6, and 7 the insights gained will be applied to fundamental processes of our planet and universe.

2
The Science of Heat and Work: Classical Thermodynamics

This chapter forms the heart of the book in that it explains the principles of classical thermodynamics while making links to social developments. We will learn about definitions of temperature, energy, and entropy. It is probably the most difficult chapter as well, because it is rather theoretical but at the same time crucial to understanding what the concept of entropy is really all about.

Historical context

Since the days of the ancient Greeks, the movements of planets and stars have fascinated humankind. Aristotle proposed one of the first models for the universe, which assumed that earth was the center of the orbits of the sun and all other planets. This model led to very complicated orbit descriptions. It was not until 1543 that a Polish monk, Nicolaus Copernicus, came up with a revolutionary idea: the sun was the center of all planetary orbits, including the earth's; this simplified orbit description tremendously. Somewhat later, Johannes Kepler (1571-1630) found, after many observations, that the orbits were not circular but had an elliptical shape. Around 1600, optics was developed so that telescopes became available. Galileo Galilei was among the first to use these new tools to study the planets. One of his biggest discoveries was a moon orbiting the planet Jupiter. Galileo also did many experiments to understand falling objects and gravitational force. Finally, Sir Isaac Newton[8] in 1687 developed a brilliant unification of orbital theories, which comprised what is now called classical mechanics. Newton showed that the movements of the planets and the falling of objects to the earth were due to the same mechanism: gravity.

Although Newton's mechanical theory was very successful in explaining many observations, it could not explain certain other phenomena, such as why heat flows only from warmer to colder areas, a commonly observed occurrence. While the modern reader may wonder at the importance of this question, it nevertheless presented a quite *fundamental* problem to science. This is because the laws of mechanics, as unified by Newton, are bi-directional. That is, moving bodies can always be stopped and then set in motion in the opposite direction, thus allowing processes to be bi-directional as well.

However, while Newton and his predecessors were developing mechanical theory, new social and economic developments were underway. Despite extensive use for thousands of years, around 1700 wood ceased to be the number one source of heat for humanity. Since about 1600, labor forces were shifting from agriculture in rural areas toward factories in and near cities. This was especially true in England, where this process moved more quickly than in the rest of Europe and reached its climax in the period from 1750 to 1830, creating the *Industrial Revolution*. Because of the strong growth in population[9], wood started to become scarce and European society moved toward coal as the main source of heat.

Coal had to be mined, and one problem with coal mines was that they would fill up with water if not continuously drained. It was to address this issue that in England in 1698, Thomas Savery filed a patent on a crude steam engine which relied on the pressure of steam to do its work. This first design was improved over time by others, including the

---

[8] Isaac Newton, who lived from 1642 till 1727, was a very talented man who was able to unify all the different mechanical laws. He also developed the mathematical tools such as differential equations which were needed in the mechanical theory.

[9] Other factors causing a scarcity of wood also played a part. The most important one was the increasing production of glass and later on the production of copper and iron, all of which needed high temperature furnaces [Burke, 1978].

2: The Science of Heat and Work

English blacksmith, Thomas Newcomen in 1712, and James Watt with his important modifications in 1765. The improvements of both Newcomen and Watt used steam pressure that was only slightly higher than atmospheric pressure (important for preventing explosions!) and condensed the steam in a cold location (condenser) of the engine, thereby creating a vacuum which did the actual work. In Figures 2.1 and 2.2, we can see the operating principle of the early steam engine, and Watt's improved version with a condenser.

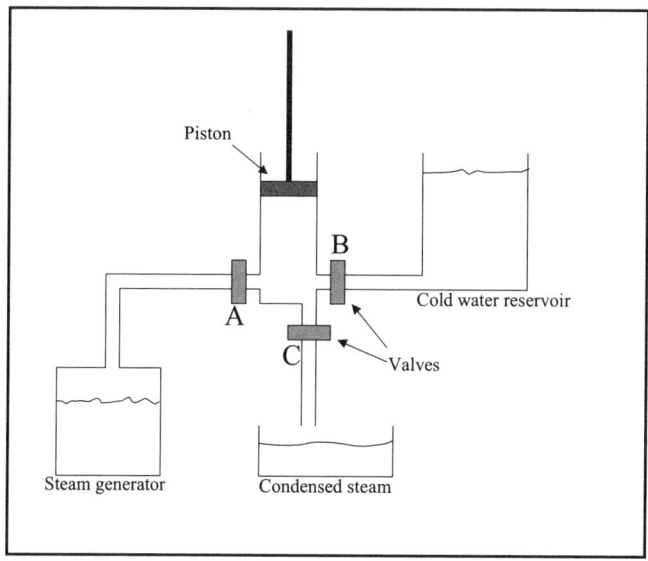

Figure 2.1 A schematic sketch of an early steam engine.

In the early steam engine, steam is generated in a vessel heated by a coal fire. When the piston is at its lowest point in the cylinder, valve A opens while valves B and C are closed. The pressure of the steam forces the piston to move up. When the piston is at its highest point, valve A closes and valve B opens. Cold water is now sprayed through a nozzle into the cylinder, and the steam condenses rapidly. This condensation creates very low pressure in the cylinder, causing the piston to come down again. When the piston is at its lowest point, valve B closes and valve C opens to let the water condensed from the steam flow into the reservoir. At that point, valve C closes and the cycle starts all over again.

A big disadvantage of this design was that the cylinder was cooled continuously, causing premature steam condensation and making the system very inefficient. However, Watt made a big improvement by introducing the condenser, whose principle of operation can be seen in Figure 2.2. The cycle begins with the opening of valve A. The condenser is kept at low temperature by the cooling reservoir. When the cylinder is at its highest point, valve A closes and valve C opens, which causes the steam to rush into the condenser, which is kept at low temperature to cause the steam to condense rapidly. This condensation creates low pressure in the cylinder, causing the piston to move down again. The advantage in this design is that the cylinder stays hot all the time, so that the pressure of the steam will not drop while the piston is in its upwards stroke.

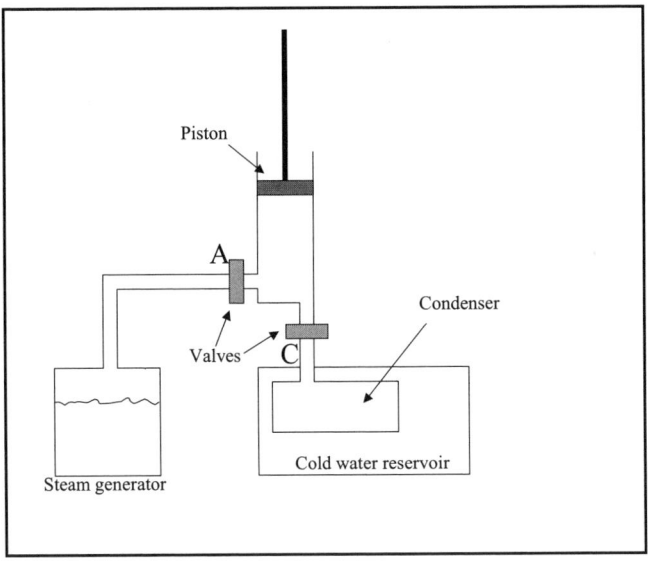

Figure 2.2 An improved steam engine equipped with a condenser. The condenser improved the efficiency of the steam engine considerably (see text for more details).

2: The Science of Heat and Work

In 1824, a French military engineer named Sadi Carnot published[10] a monograph entitled *Reflexions sur la puissance motrice du feu et sur les machines propres a developper cette puissance,* which translates as *Reflections on the motive power of heat and on machines fitted to develop this power.* Carnot tried to answer two important questions: Is there a limit to the efficiency (ratio of energy in and work out) of the steam engine[11], and are there other media than steam to produce power? By carefully analyzing the heat and temperature characteristics of the steam engine, Carnot concluded that the efficiency of the engine depended only on the temperature *difference* of the hot and cold areas of the engine. This seemingly simple conclusion birthed a completely new physical theory called *thermodynamics*. It also marked the first time that science had been put to work for an economic reason, to improve the performance of an important commercial machine[12]. (Previously, scientists had studied predominantly "elevated" issues, such as planetary orbits.) Interestingly, Carnot came to his conclusions well before the law of conservation of energy (i.e., the First Law of Thermodynamics) was known and generally accepted. Although historians disagree on the impact of Carnot's work on the development of thermodynamics, it's revealing that Sir William Thomson[13] in 1849 adapted Carnot's ideas to reflect new insights into the nature of heat.

---

[10] There has been some confusion what the exact time of publication was (Epstein, 1945). Fact of the matter is that the work was published on 12 June 1824 and Pierre Girard already discussed it in the Academie des Sciences in Paris on 26 July of that year.

[11] This was a very relevant question at that time. Since Newcomen's steam tool the efficiency was improved dramatically from about 0.5% for Newcomen's tool to about 15% for the tools in operation at Carnot's time. No clear limit to the efficiency was visible; for more details see the section on the Carnot cycle.

[12] Interesting is that these developments and innovations in physics in the mid 19th century sparked near the end of the 19th century a new approach in economics called neoclassical economics. One important driving force was the increased understanding in physics of the concept of the relation between heat and energy and the conservation of energy. Another one was the introduction in economics of the rigorous mathematical framework used in physics which led to new mathematical descriptions of the economic process, something that was not appreciated by all economists at that time of which many had no mathematical background at all [Mirowski, 1951]

[13] Sir William Thomson, better known as Lord Kelvin and a well respected scientist, introduced the absolute temperature scale named after him in 1848.

It took quite some time before it was generally understood that heat and mechanical energy are equivalent, and can be converted back and forth. In 1693, Leibniz showed that in an isolated mechanical system, the sum of kinetic energy (called *vis viva*, or life force) and potential energy (*vis mortua*, or dead force) stays constant. In 1798, Count Rumford (also known as Benjamin Thompson) showed that mechanical energy could produce heat. He observed that a lot of heat was generated in boring cannon barrels, and translated the produced heat into a measurement based on the power of a horse (thus introducing "horsepower" into the world's mechanical lexicon). In 1842, Robert Julius Mayer, a German physicist, calculated the ratio of heat and work from heat studies on gases. He came to the conclusion that the amount of heat needed to warm up one kilogram of water one degree Celsius (which is 4200 J or 1000 cal of heat) is equivalent to the amount of energy needed to lift 400 kg one meter! In 1843 and 1849, James Prescott Joule calculated from studies of churning water and other liquids with paddle wheels that one calorie of heat is equivalent to 4.18 "Joules" of mechanical energy[14]. Additional work by Hermann von Helmholtz eventually led to the formulation and acceptance of the law of conservation of energy[15]. However, people still believed at the time that heat was a kind of fluid called *caloris* [See Mendoza, 1961]. This idea resulted in a law of conservation of heat, which was used for some time before being discarded. (In Chapter 3 we will see how heat and temperature can be much better related to matter by analysis at the atomic level.)

Now we come back to the previously discussed riddle of why heat flows only from hot to cold regions. A leap in understanding this phenomenon was made in 1865 by German scientist Rudolf Clausius. Clausius realized that heat (or energy), when used to produce work, is

---

[14] The joule will be the unit we will use for energy in this book; one cal is 4.184 J.

[15] Hermann von Helmholtz (1821-1888) reported on July 23 in 1847 on the principle of conservation of energy and showed that he had acquired a deep understanding of this principle. He was, together with Rudolf Clausius, the founder of what was called the Berlin School of Thermodynamics where he succeeded Magnus as the director of the Physical Institute. The influence of this school on the development of thermodynamics was crucial. It is almost unbelievable how many famous scientists were connected to this school. To name a few: Walter Nernst, Max Planck, Albert Einstein, Erwin Schrödinger and Leo Szilard. See for more details [Ebeling and Hoffman, 1991].

partly converted into a "non-available" form and cannot be used to generate power. He introduced the concept *entropy* (S), which combined the German word *energie* (energy) with the Greek word *trope* (turn). Further observations by several physicists revealed that in spontaneous processes, entropy always increases and never decreases, and that the product of the temperature (T) and the change in entropy (ΔS [16]) equals the amount of heat or energy that is converted into non-available form. This work concluded that energy has two states: free or available, and bound or latent. Free energy always goes from free status to latent status, and all energy eventually becomes latent. This state, at the cosmologic level, is called Heat Death (more on that later).

Paul Epstein[17] has noted that the real breakthroughs in the development of thermodynamics came not from physicists, but from medical professionals such as Clausius, Mayer, and Helmholtz. The medical profession's interest in heat phenomena undoubtedly was fueled by the ongoing debate over the origin of animal heat. Around 1820, it was thought that about 80% of animal heat was created from the combustion of carbon and hydrogen, while the remaining 20% was believed to come from the internal friction of blood moving through the veins; some support for the latter assertion could be found in Count Rumford's cannon boring experiments, described above.

How easy is it to go back?

Imagine we take a walk in a forest and follow a dry, level trail that allows us to turn at any point and go back the way we came; you could call this walk a "reversible process." However, if the trail led to a steep canyon that we had to jump in, it probably would not be as easy to return to our starting point along the same path; this walk could be called "irreversible." Analogous situations can be found with chemical or physical processes. Consider, for instance, an ice cube in water. If both cube and water are at 0°C, we can melt the cube by adding heat or grow

---

[16] Recall that the symbol Δ is used to indicate a *finite* change in a parameter, in this case the entropy. ΔS indicates thus a finite change in entropy

[17] Paul Epstein, professor of theoretical physics at the California Institute of Technology, was a student of Arnold Sommerfeld. Epstein wrote a textbook on thermodynamics.

the cube by extracting heat. In other words, the process can go easily in two directions; another way to say this is that the melting occurs under reversible conditions. However, if the water surrounding the cube (still at 0°C) has a temperature of 10°C, we won't be able to reverse the melting process easily, since we first would have to cool the water down to 0°C before the freezing process could start – and so this process is irreversible. In a nutshell, "reversible" means we can reverse process at any moment and go back to the starting point along the same path as we came.

So why are we discussing reversibility? First, the only way we can determine the change of entropy of a process is if that process follows a reversible path. I realize that this is not an obvious statement at this time. However, we will see that we can calculate the entropy change for a given process from the heat changes that occur, but *this can only be done when we have a reversible route from the beginning to the end* conditions of that process. For any given irreversible process one can always design an alternative reversible route, and since the entropy change depends only on the *beginning and end* conditions, this method is often used to calculate the entropy change[18]. Second, spontaneously (naturally) occurring processes proceed typically under irreversible conditions (including the chemical processes in living beings), which greatly impacts process' efficiency in that reversible processes are the most efficient but also the slowest. In fact, the only place where we come close to having reversible conditions is the laboratory test bench! For these reasons, we need to understand better the nature of reversibility, or more accurately, the lack of it.

To give you a better feel for what's going on, let's do a thought experiment: we inject a gas at low pressure into a chamber equipped with a moving piston and probes for measuring the pressure, temperature, and volume of the gas. The entire chamber is immersed in a big reservoir

---

[18] I admit that we run a bit ahead of ourselves here. After reading further in this chapter, however, we will understand the relation between the calculation of entropy, heat, and reversibility much better. Also, it is important to understand that energy and entropy are state functions, which means that they are determined only by the conditions (such as pressure and temperature) of a given state and absolutely *not* by the path that led to that state. This is not true for work and heat, they *do* depend very heavily on which way is followed to arrive at a given state.

filled with water that is maintained at a fixed temperature $T_{reservoir}$ (see Figure 2.3).

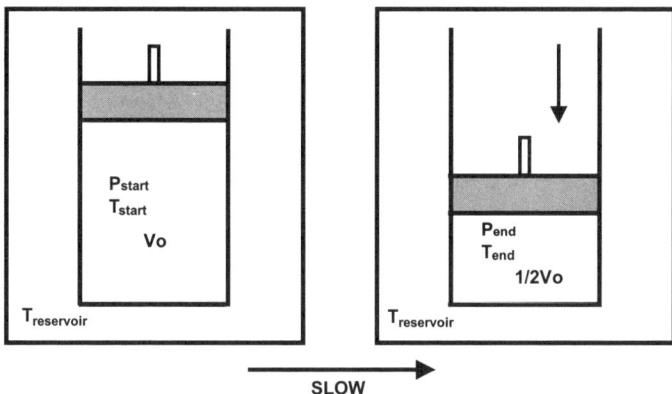

Figure 2.3 By very slowly pushing the piston down we can change the volume of the enclosed gas to, in this case, half its original volume.

Thermodynamically, the state of a gas can be described by three parameters: pressure (P), volume (V), and temperature (T). This assumes, however, that in the system (chamber) under consideration at any point in that system we have the same pressure and temperature. Another way of expressing that is to say that the system is in equilibrium: when left alone it will not change anymore. This implies that T_{start} (the initial temperature of the gas) will be identical to $T_{reservoir}$ of the reservoir.

Now, we begin to reduce the chamber's volume to half its original volume. We do this very slowly (say, taking an hour) so that at any moment, pressure and temperature are uniform throughout the chamber (no shockwaves, turbulence, and that kind of mechanism to occur). We also continuously and simultaneously record pressure, volume, and temperature, as in Figure 2.4. Importantly, we observe that the temperature of the gas does not change during compression, but stays equal to the temperature of the reservoir at $T_{reservoir}$ (thus $T_{reservoir} = T_{start} = T_{end}$). This is because we went very slowly and there was plenty of time for the heat to exchange and to keep gas and reservoir temperatures the same.

Next, we reverse the process by expanding the chamber to its original volume – again, very slowly; the pressure goes from P_{end} to P_{start}. Again, we carefully record the pressure, volume, and temperature of the gas in the chamber. If we do the entire process slowly enough, we see that pressure and temperature pairs are exactly the same for both forward and backward processes. These, of course, are reversible changes because we can go in both directions easily return to our starting point (P_{start} and T_{start})

Now let's assume that we start with the same conditions as above, but that this time we compress the volume very quickly (say within a second our so, see Figure 2.4). Again we record pressure, volume, and temperature during this process. This time, we see that the pressure and temperature values differ markedly from those obtained in the reversible process. Moreover, we notice that even after we have completed the volume reduction, pressure and temperature are not constant for some time. What happened is that the pressure in the chamber varied during the expansion, as well as the temperature throughout the chamber. Because of the non-uniform temperature and pressure distribution, convection (or "wind") occurred in the chamber. If we wait long enough after the completion of the compression, we see that the temperature and pressure assume the same values P_{end} and $T_{reservoir}$ as in the reversible experiment. This is of course because the gas has no choice: the temperature at equilibrium will that of the reservoir and the volume is back to its initial value, therefore the pressure *must* be the same as the initial value as well.

Now, suppose we expand the gas to its original volume, again very quickly. We will see that the pressure and temperatures recorded will not follow the same path as during the compression, nor the paths as seen during the reversible compression. Clearly, the process is not able to pass through the same values for pressure, volume, and temperature combinations going forward as going backward. This type of process, therefore, is called irreversible: we cannot go back at any point and reach the same pressure and temperature combinations (or states) as we did going forward.

2: The Science of Heat and Work 19

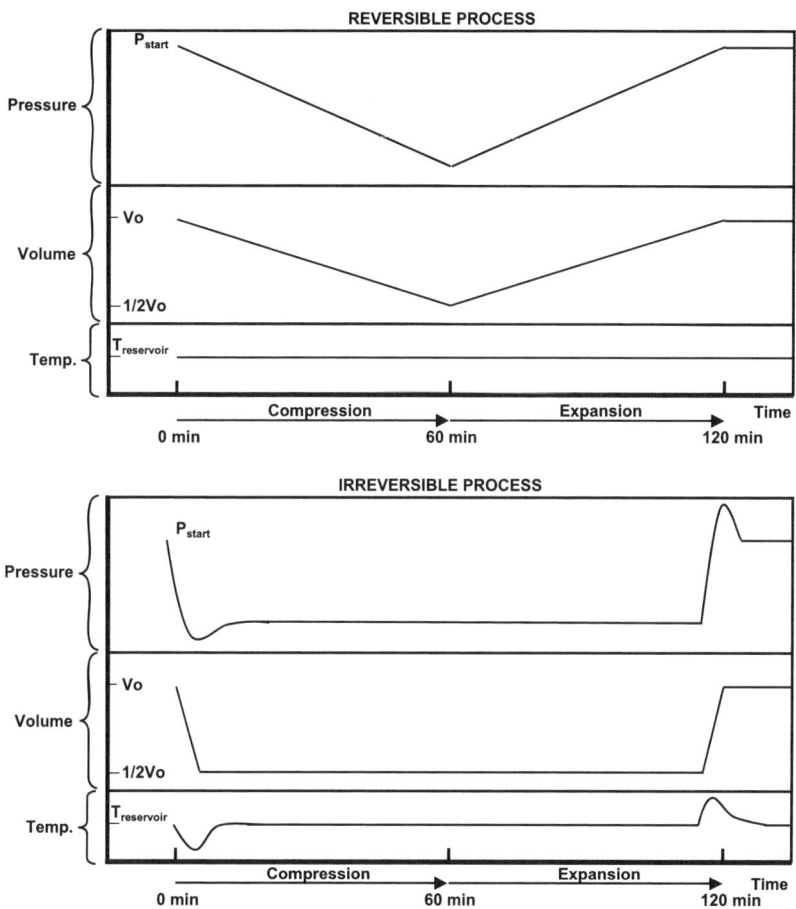

Figure 2.4 This is how the output of the multi-pen recorder would look in case of a reversible (top) and irreversible (bottom) expansion and compression of a volume of gas in a container immersed in an infinite reservoir at fixed temperature.

In fact, classical thermodynamics only studies equilibrium states and reversible state transfers. Although spontaneous processes which occur in nature are almost always irreversible, it is still possible to draw important conclusions for irreversible processes from certain reversible studies, as we will see later.

Heat, energy, and mechanical work

What do you mean when you say it's warm in here?

Before we go further with our discussion of the relation among energy, heat, and work (which will lead to the introduction of the concept of entropy), we must say a few words about temperature. Temperature is the most important parameter in thermodynamics. That's because it determines the energy of a gaseous system, and if we track temperature over time in a given system, we'll be able to tell whether that system is in equilibrium. However, determining temperature is not a trivial matter.

Of course, we all can feel temperature differences. Imagine you're driving your car, and suddenly your exhaust pipe falls off. Your first reaction is probably unprintable, but your second likely will be that you stop, walk back to retrieve your tail pipe, and again utter something unprintable. That's because you dropped the pipe when you felt how hot it was. From such a real-world experience, measuring the difference between hot and cold seems straightforward – but in thermodynamics, it's not that simple a task. Stay with me a moment, and I'll explain some more.

Suppose you come from freezing weather into a room, which is maintained at 15°C (or 59°F). Initially you will think the room is warm, but after a while, you'll start to feel a bit chilly. From this example, and your experience with the tail pipe, we learn that we're good at determining temperature differences, but not so accurate about knowing the absolute temperature. So it's not surprising that the determination of temperature has received considerable attention in the development of science. To illustrate, let's look at the history of how thermometers were developed [Moore, 1972].

In 1631 a French physician, Jean Rey, monitored the temperature of his patients using a glass bulb with a stem partially filled with water. In 1641, an alcohol in a glass thermoscope was invented by Ferdinand II, Grand Duke of Tuscany. Calibration of the thermoscope was done by taking the coldest temperature seen in the winter and the warmest in the summer. Dalencé improved this calibration in 1688 by ascribing −10° to the melting of snow and +10° to the melting of butter. In 1694, it was Renaldi who proposed using the melting point of water and the boiling point of water as the lower and higher points of the scale on the thermometer.

Table 2.1 Comparison of three common temperature scales		
K (Kelvin)	C (Celsius)	F (Fahrenheit)
0	-273	-460
273	0	32
293	20	70
373	100	212

Today, the two best-known temperature scales are Fahrenheit and Celsius (see Table 2.1). In scientific discussions, these two scales are not very practical because their zero points are not the lowest possible temperatures in nature; in both systems, temperatures can become negative. Also, both scales are based on rather subjective criteria: Celsius assigns 100 degrees between the temperatures of boiling water and of melting ice (both at one atmosphere of pressure) as suggested by Elvius in 1710 and used by Celsius, who was a Swedish astronomer. The scale of Fahrenheit is based on a similar arbitrary interval assignment (32 degrees for freezing water and 212 for boiling water).

Around 1805, French scientist Joseph-Louis Gay-Lussac was investigating the relationship of gas volumes to temperature at constant pressure. First, he determined the volume of an enclosed gas at 0°C, and called this V_o. Then he varied the temperature of the gas while keeping the pressure constant (by adapting the volume of the enclosure). In Figure 2.5, we can see his results.

Clearly, as we continue to lower the temperature, the volume of the gas decreases. Moreover, there will come a point when the volume of the gas will approach zero. At the time Gay-Lussac was doing his experiments, it was impossible to obtain temperatures that low, but it was possible to extrapolate the curve to zero volume. It was found that that condition happened at about −273°C.

In 1848, drawing from the work of Gay-Lussac and others, the British physicist Lord Kelvin proposed a scale based on this concept of zero gas volume, and called it absolute zero. Thus we got the Kelvin scale that has no temperature below zero degrees[19] -- and that's the scale we'll use for the remainder of this book, unless otherwise indicated.

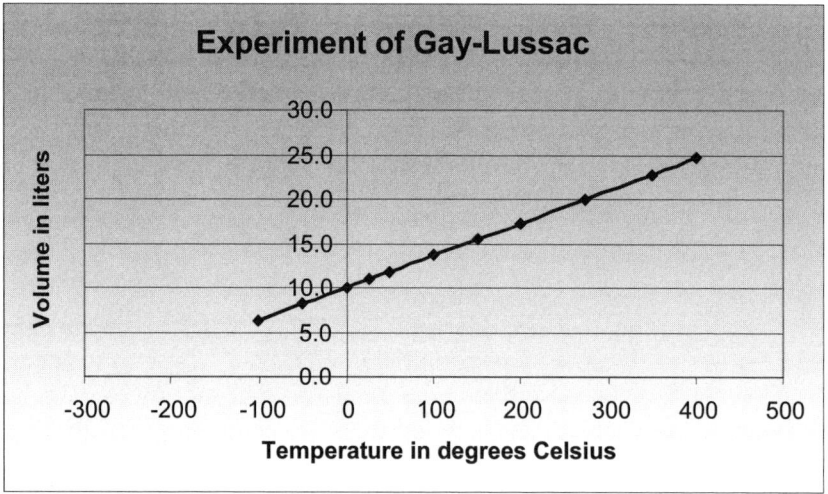

Figure 2.5 Gay-Lussac extrapolated the volume of a gas versus the temperature and was in that way able to determine an absolute point of zero for the temperature.

Temperatures can be measured in numerous ways. The best known methods use the temperature-induced changes in volume of a liquid, such as alcohol or mercury. However, many other methods are available, such as gas pressure or the electrical resistance of a wire.

Thermometers work because of a very important phenomenon: when a hot and a cold body are brought in contact, heat flows from the hot to the cold body.[20] Eventually, the two bodies reach the same temperature, or as scientists like to say, they attain thermal equilibrium. Thus, the longer a thermometer remains in contact with a person, the

---

[19] In Chapter 3 we will get another insight into what the fundamental reason is behind the absolute temperature.

[20] Although that appears to be a trivial statement, later in the book during the discussion of the concept of entropy we will see that it is not trivial at all! To explain why heat does flow spontaneously from hot to cold is not that straightforward.

more accurate[21] its reading will be. Certainly, the speed of the reading also depends on position: an electronic thermometer placed under the patient's arm will take about two minutes to start beeping, but a thermometer under the tongue will be ready much faster –because the thermal contact in the latter case is much better.

What is heat, what is work, and what is energy?

There are three very important concepts in thermodynamics, the science that studies the transformations of energy: These are energy, work, and heat. Everybody has a gut feeling for what heat, energy, and work are. But few of us can define exactly what those words mean. We shouldn't be embarrassed, though – it took scientists several hundred years to reach a consensus on those topics. To understand these phenomena, we must step beyond their everyday meanings and use the more precise language of thermodynamics. The definitions of Peter Atkins [Atkins, 1983] are simple and straightforward: *Work* is done if one changes the height of a weight. *Energy* is the capacity of a system to perform work. *Heat* is the transfer of thermal energy between two bodies that are at different temperatures; as long as these bodies remain in contact, the transfer will continue until the temperature difference is zero. The flow of heat will change the amount of energy in each body, by redistributing energy throughout the system (i.e., the two bodies), or by exporting or importing energy. Thus, the energy content of a system can change not only by using the energy to perform work, but also because heat can flow from that system to the environment, or vice versa.

The term *energy* probably was used first by d'Alembert in the French *Encyclopèdie* of 1785: "there is in a body in movement an effort of *energy* which is not at all in a body at rest" [Moore, 1972]. Most of us know from high school that the mechanical energy of a given system (say a football thrown by a quarterback) has two components: kinetic energy (which is determined by the velocity (v) and the mass (m) of the

---

[21] Accurate is relative here. In order for the thermometer to work, heat must be exchanged between the thermometer and the body we want to know the temperature of. This means that the temperature of the body will drop (in case of body temperature higher than that of the thermometer) while the temperature of the thermometer will rise. In most situations the change in body temperature is small.

football, and expressed as $1/2\ mv^2$) and potential energy (which is determined by how high the ball is above ground level, and expressed as h times g times m, where g is the acceleration of the gravitational field). Thus for instance, a ball at rest on a roof will have zero kinetic energy, but a certain amount of potential energy – because it can always fall off the roof. On the other hand, a ball on the ground at rest has neither kinetic nor potential energy. If we kick the grounded ball and give it a certain velocity, it will have kinetic energy but still will lack potential energy.

The message is clear: energy comes in many different combinations. Importantly for us, Newton's classical mechanics revealed that potential energy can be converted to kinetic energy, and *vice versa* Even more significant, Newton found that the sum of kinetic and potential energy is constant. This means that a ball dropped from the roof will gain kinetic energy as it speeds up, but at the same time will lose potential energy as it loses height; again, the sum of kinetic and potential energy remains constant. This observation is commonly known as the Law of Conservation of Energy.

We can take any system that has energy and use it for practical purposes, such as generating heat or work. This involves the transfer of energy from one location to another, or sometimes from one system to another. Transferring energy is when things become really interesting -- as we will see below.

We have seen above that scientists in classical mechanics came to the conclusion that the sum of kinetic and potential energy must be constant. Thus from an energetic point of view, kinetic and potential energy are equivalent, and are just different forms of energy. In thermodynamics we have an analogous situation pertaining to the relationship between heat and work. We have seen that heat can be converted into work, or vice versa. Experiments done by Benjamin Thompson[22] during the boring of cannon barrels showed that mechanical work produces heat. In many cases, this transfer of mechanical work into heat causes trouble in daily life: for instance, a lot of cooling must be applied to the brakes of our automobiles to prevent them from getting overheated. In coming to a stop, a 1500 kg car traveling at 100 km an hour will produce about 600,000 J of heat through its brakes. That's

---

[22] Benjamin Thompson, also known as Count Rumford, noticed that a lot of heat was produced during the boring of cannon barrels in Munich around 1800.

2: The Science of Heat and Work

enough heat to bring almost 1.5 liters of water from 0°C to boiling point within a few seconds[23]!

The opposite effect — the conversion of heat into mechanical work — proved much more important to economics, and was first applied in steam engines. The principle behind this event also led to a reformulation of the principle of conservation of energy as the First Law of Thermodynamics.

The law of conservation of energy is in fact a postulate, which means it is based on many observations but no proof is given. To illustrate, suppose we throw a stone into the air 1,000 times, and we see that the stone always falls back to earth. We can now formulate a postulate, or law, that heavy things thrown aloft will always fall back. We have no model or theory that explains this, but the postulate is very accurate in making predictions. The same is true for the First Law of Thermodynamics. We don't know why the energy is conserved, but no evidence exists that there has ever been any deviation from that law. (However, many contrarians have performed experiments and built "perpetual motion machines" that allegedly defeated the First Law; as described later.)

The First Law can be stated simply as:

$$\Delta U = \Delta Q + \Delta W \qquad (2.1)$$

This formula says that a change in energy (ΔU)[24] can be achieved in two ways: through a heat transfer (ΔQ) or through an amount of work done by or on the system (ΔW), or by a combination of both. Now we can start to understand better the relationship among energy, heat, and work. Energy can be added to, or taken from, a given system by transferring heat or/and work. Most of the time, heat transfer is shown by a change in

---

[23] To put it differently this is roughly 6% of the amount of energy an adult needs to take up everyday!

[24] In the literature both the letter U and E are used for energy. For now we will use the letter U. In thermodynamics typically the *internal* energy is considered, which is the sum of the kinetic energy and the potential energy. Nuclear energy or potential energy because of the location of the system in an electrical or gravitational field is excluded. For simplicity we assume for now that changes in internal energy occur only because of changes in heat, which is throughout the book adequate to illustrate the concepts. Later in the book we will expand on this.

temperature (but not always, as we will see below), while work mostly is visible in some form of motion. (Indeed, several other forms of work are possible: passage of electrical current within a motor, magnetic cooling (to achieve very low temperatures), and nuclear fission, for example. However, for now we will consider only work done by the expansion or compression of a gas. It can be easily shown that if a container of gas changes volume with an amount ΔV while keeping the pressure constant, the amount of work done is $P\Delta V$.

Before we go on, it's important to realize that heat and work should *not* be considered different forms of energy; both are *energy transfers* between systems. Thus, it's incorrect to speak of "the heat or work of a system" but correct to refer to the *energy* of a system. [25] The scheme below shows the different ways that energy can be transferred.

Different ways to transfer Energy from one region to another

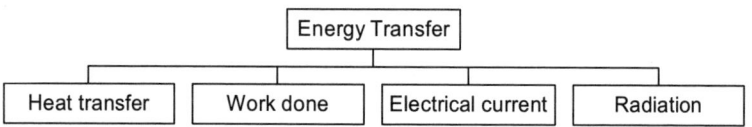

To bring this back to everyday life, here are a few examples of the amounts of energy transfers involved in familiar situations:

- Heat needed to make a cup of coffee: 15 kJ (heat from the fire transferred to the water)
- Work required to take the stairs up one floor: 2500 J (chemical energy in the muscles converted into work)
- Each of your heartbeats: about 1 J
- Energy we get from eating three meals a day: about 10,000,000 J (chemical energy present in the food is stored in the body in a suitable form such as fat and glycogen)

---

[25] Let me expand a bit on this statement. The energy content of a system in a certain state depends only on the relevant parameters of that state and not on the way the system was brought to that state. For example, the energy of a gas can be shown to depend in good approximation only on the temperature.

2: The Science of Heat and Work

The formulation of the First Law, as shown in formula (2.1) above, deserves more study to deepen our insight. Let's consider a few points:

a) It might not be immediately clear how this formula represents the First Law. However, it can be proven mathematically that ΔU depends *only* on the conditions (pressure, temperature, and volume) of the start and finish situations, and *not* on the path we took to get from beginning to end state [26]. In scientific language, the energy involved is a *state function*. On the other hand, the change in heat and work, ΔQ and ΔW, absolutely depend on which path is followed to get from start to finish, however, their *sum* does not. Thus, if we have two routes (A and B) from start to finish, then in general ΔQ_A is not equal to ΔQ_B, and ΔW_A is not equal to ΔW_B, but $\Delta Q_A + \Delta W_A = \Delta Q_B + \Delta W_B$, since the change in energy is independent of the path followed. Now, suppose we have a cyclic process and thus start and finish conditions are identical. In this case, $\Delta U = 0$, which is another formulation of the law of conservation of energy. In a cyclic process there still can be an exchange of heat and work, but after each cycle the amount of work done will be equal to the amount of heat exchanged, or $\Delta Q = -\Delta W$. It is also clear that in a non-cyclic process, any change in ΔU must be balanced by a similar amount $(-\Delta U)$ elsewhere, since the total energy must stay constant[27]. For instance, if you raise this book one meter, the book will gain 15 J in potential energy, but the work needed from your muscles will require 15 J of energy, which must come from metabolic burning of some of the food you've eaten.

b) Gases at lower pressures have an important property which makes the discussion here much easier: in general, the energy in gaseous systems depends *only* on temperature, and not on pressure or volume. (Gases at low pressure that exhibit this property are also called "ideal"

---

[26] A state points to a specific situation of the system described, in this case, by pressure and temperature. This path independency is quite crucial!

[27] This is related to the determination of a system. You can always define a system such that ΔU is zero; it depends how large you make your system. Worst case your system needs to be as large as the universe. When you start with the universe and make your system smaller and smaller there will be a point that ΔU for your system is no longer zero meaning that there is an exchange of energy between your (sub) system and the environment (the rest of the system).

gases[28].) It is an approximation of the real situation but works rather well in a number of cases. In the language of physics, this means that the energy of a volume of gas is a "state function," since that energy depends only on temperature, regardless how we arrived at the temperature.

c) Whenever we have a process where we keep the temperature constant (called an *isothermal* process), it can be proven that for an *ideal* gas, the energy change is zero. This is another way of saying that the energy of an ideal gas depends only on the temperature. Again, this is a fact that is very often used in thermodynamic analysis. Thus for an isothermal process, we can write that $\Delta U = 0$ and $\Delta Q = -\Delta W$.

d) Another important process type is where we completely isolate the system from the environment, so that no heat exchange is possible. Such a process is called *adiabatic,* which can be written $\Delta Q = 0$. Thus in an adiabatic process, the energy change equals the amount of work done, since $\Delta U = \Delta W$.

We will see in the next section that we can apply the principles as explained under a, b, c, and d directly to the analysis of what is often called the Carnot cycle. The cycle is named for the work of Carnot, upon which Clausius later founded his proposal for entropy, and we will examine it in much detail below. Even if you find the prospect of this topic a little dry, hang on – I guarantee that you'll find an exotic and entrancing world waiting for you!

Entropy and the Second Law of Thermodynamics

The Carnot cycle

When an instructor has to teach a difficult subject, such as a foreign language, what is the likely approach? Most teachers start with some simple words and sentences, and slowly build on that base to increase the level of complexity. In science, a similar approach is often followed. First, a simple model is built to approximate the actual situation so that a mathematical analysis can be performed. However, initial results from such a model are approximate or valid over a limited range. Frequently, further refinements are then applied which typically make the model and

---

[28] This is a very crude definition of an ideal gas but good enough for our purpose.

analysis more complex, but which yield better approximations of the real system under consideration.

As mentioned earlier, Carnot was studying the efficiency of steam engines. The purpose of a steam engine was to convert heat from the combustion of coal into work. In the steam engine, the energy transporting medium is steam (water vapor). But in order for a steam engine (or any engine) to be practical, it must be cyclical – that is, the piston producing labor must come back to its original position after every cycle. As was pointed out earlier, the efficiency of the steam engine in terms of input of heat by burning coal and output of mechanical work was improved considerably over the years since Newcomen. Watt improved the efficiency by about 25% over Newcomen's steam engine by adding the condenser. Further improvements were achieved by alternating steam pressure to both sides of the piston, and by converting the linear movement of the piston into a rotary motion. Then in 1781, Jonathan Hornblow invented a "compound engine" that used two cylinders. The second cylinder ran on steam that escaped from the first cylinder, expanding engine power. This concept was revitalized by Arthur Woolf in 1804, raising efficiency by 7.5%. Richard Trevithick then introduced high-pressure steam to the Watt engine, again increasing efficiency by about 15%.

This led people to believe that there was no limit to engine efficiency[29]. A young French engineer, Sadi Carnot, undertook a study to understand better the fundamentals of the steam engine. Because the analysis of a real steam engine would have been rather complex, he designed a model system that allowed him to use rather simple arguments to analyze the heat and work exchanges. His model of steam engines is very famous and is called the Carnot cycle. Fortunately, the results he obtained from this model were quite generally valid and improved insight into the working of steam engines. We will now

---

[29] A similar situation is found in the development of the optical microscope. The first work was done around 1680 by Anton van Leeuwenhoek, a Dutch scientist, who was able to achieve a magnification of 275X using a single handcrafted lens. Fraunhofer improved this magnification considerably by putting two lenses in series. At that point, the thinking was that there was no limit to the achievable magnification, since one, in theory, could put an unlimited number of lenses in series. However, Ernst Abbe discovered and explained that the usable magnification was limited by the wavelength of the used light, which determines the achievable resolution.

describe in detail how the Carnot cycle works, because it will lead to a better understanding of the concept of entropy.

We will start again with a system that consists of a cylinder filled with a gas and equipped with a piston. The cylinder is immersed in a reservoir, which keeps the temperature constant. In addition, we can if necessary make the walls of the cylinder impervious to heat through some kind of isolation.[30] We start our process at a pressure P_1 and at a temperature T_{high} and volume V_1 (see also Figure 2.6). Suppose we allow the gas to expand *reversibly*[31] and *isothermally* (thus at constant temperature T_{high}) to a volume V_2 by going from state 1 to 2. The pressure, of course, will drop and will assume a value of P_2. Now, let's do a first analysis. As mentioned previously, the energy of the system will not have changed, since for an ideal gas the energy depends only on the temperature and thus $\Delta Q_{1,2} = -\Delta W_{1,2}$. In other words, the amount of heat taken up from the reservoir at temperature T_{high} is *completely* converted into work.

This is a nice situation and if we only had infinitely long cylinders, we could make very efficient engines. But we know that cylinders don't come in infinite lengths, and so we need a cyclic kind of engine. One possibility is simply to let the piston go back from volume V_2 to the start volume V_1. Again, if we do this under reversible and isothermal conditions, we will end up at exactly the starting conditions P_1, V_1, and T_{high}. We thus would give back all the heat we took in the first step, and the net amount of work produced would be zero. Therefore, in order to have an engine which produces a *net* amount of work, we need to do something more: we have to make sure that the return step to the initial starting position includes a step at a lower temperature!

This statement might be puzzling. (Why the need for a lower temperature?) But stay with me for a moment. One simple way to go to a lower temperature is to thermally isolate the system from the bath and expand the volume of the cylinder to a volume V_3 under *reversible* and *adiabatic* (I recall here that adiabatic means no exchange of heat and thus $\Delta Q_{2,3}= 0$) conditions. The temperature and the pressure in the cylinder will then drop to T_{low} and P_3. At this point, we carry the cylinder over to

---

[30] That's the nice thing about theoretical experiments. All kind of conditions can be imposed without us needing to worry about how exactly to do all of that.

[31] Reasons to impose reversibility at this stage are given later in this chapter.

another bath at temperature T_{low} and remove the heat-insulating jacket [32]. In order to make the process cyclic, we want to return to the initial values of pressure, volume, and temperature. We can do this with two more steps. The third step is to decrease *isothermally* the volume of the gas to V_4. Of course, pressure increases but temperature remains at T_{low}. The fourth and last step is that we again apply the heat-insulating jacket and decrease the volume *adiabatic* from V_4 to V_1. The pressure will come back to P_1 as will the temperature from T_{low} to T_{high}.

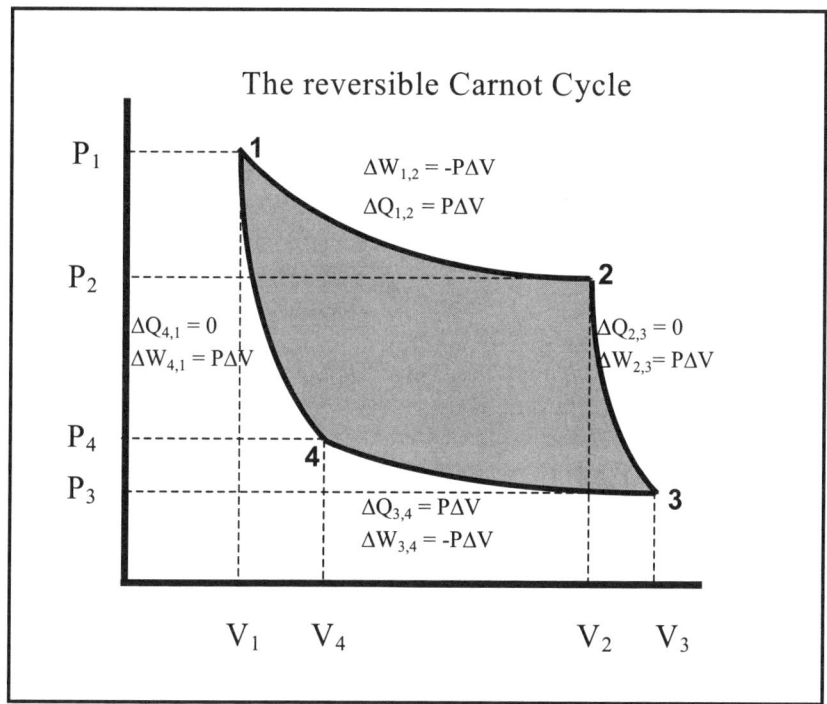

Figure 2.6 A reversible Carnot cycle going from 1 to 2 to 3 to 4.

Will this reversible and cyclic Carnot process actually give us a net amount of work we can use to drive a wheel or something similar? The answer is yes. Mathematically, it can be shown that the net amount

---

[32] Keep in mind that we deal here with a model only, and that allows us to do these impracticable manipulations.

of work is given by the shaded enclosed area [33] in Figure 2.6. This is an important result. We immediately can see which parameters we have to change to get more work out of one cycle. The surface enclosed will become larger for instance, if we make the pressure difference between P_1 and P_3 larger. This of course is determined by the two temperatures T_{high} and T_{low}. Thus, if we make T_{high} higher and T_{low} lower, we will get more output out of our engine for a given input quantity of heat of the high temperature reservoir. Indeed, it can be shown that the efficiency of the Carnot cycle is determined by equation (2.2):

$$Efficiency = \frac{\Delta W}{\Delta Q_{1,2}} = \frac{T_{high} - T_{low}}{T_{high}} \qquad (2.2)$$

The efficiency is defined as the ratio $\Delta W/\Delta Q_{1,2}$ or the amount of net work delivered, divided by the amount of heat taken from the hot reservoir (or the heat source, if you like). The net amount of work delivered in the cycle, ΔW, is defined by $\Delta W = \Delta W_{1,2} + \Delta W_{2,3} + \Delta W_{3,4} + \Delta W_{4,1}$. This is, of course, a very significant result. Carnot showed that there is a limit to the efficiency of an engine which transforms heat into work, and that the high and low temperature regions present in the engine determine this limit. In Figure 2.7, we can see how the efficiency ranges with T_{high} and T_{low}.

Getting confused? Well, let's compare a steam engine to a waterfall. It's obvious that the higher the waterfall, the more energy will be available to do work. Water converts potential energy into kinetic energy by falling from a high spot to a lower level. In the heat engine, we have a similar situation. Heat goes from a high temperature to a low temperature, and this conversion transforms partly in work. However, the amount of heat that is given up to the cold reservoir going from step 3 to 4 can never be used again to produce work at temperatures above T_{low}. Only if we can find another reservoir with a temperature lower than T_{low}, can some of the heat again be transformed in work.

Now you know why engineers have always wanted to increase the pressure in steam engines. While the temperature of steam at atmospheric pressure is 100°C, at many atmospheres of pressure it can easily rise to several hundred degrees Celsius (for example, at 6

---

[33] As a matter of fact this is not difficult to prove.

atmospheres the temperature of steam is about 160°C). This will increase the difference between T_{high} and T_{low} while making T_{low} as low as possible, and so we can improve the maximum achievable efficiency considerably. (This brings up an interesting fact: formula (2.2) indicates that your car will run more efficiently in winter than in summer because the temperature of the cold heat reservoir (i.e., the outside air) is lower.[34])

However, the original motivation for increasing pressure was to get more power out of steam engines.[35] Carnot's work provided fundamental insight into how to do so, which resulted in an expansion of uses for the steam engine far beyond the coal mines. For example, the proportion of ocean-going ships powered by steam went from 45% in 1890 to about 75% by 1910. [Nov. 1972].

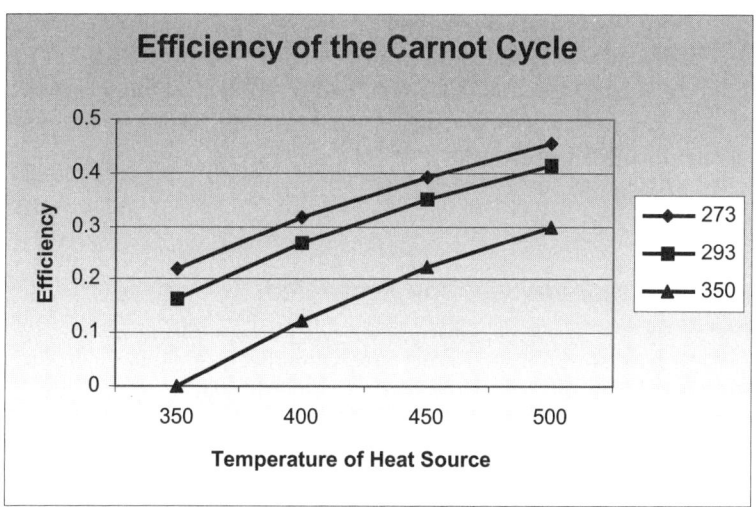

Figure 2.7 The efficiency of a Carnot heat engine at three different temperatures (273, 293 and 350 K) of the cold heat sink.

---

[34] However, another complication comes into play, and that is that the viscosity of the air also changes with temperature and with that also the friction the car feels.

[35] While James Watt improved considerably the efficiency of his steam engine, which ran at a pressure only slightly greater than one atmosphere, he opposed increasing steam pressure. He feared that higher pressure could lead to unsafe situations and accidents (which indeed happened from time to time).

By now, you may be wondering about the relationship between the Carnot cycle and a real steam engine. The answer may seem disappointing: The different stages in the steam engine are not directly related to the four process steps in the Carnot cycle, because the latter is an ideal description of a cyclic engine. Nonetheless, Carnot made sure that the essential elements of real steam engines were captured in his model: a high temperature heat source (the boiler), a low temperature heat sink (the condenser), and a cyclic process route. The power of this approach is that it is valid for all kinds of cyclic engines which convert heat into work. The Carnot cycle also works in reverse: if we go in the opposite direction (starting at state 4 and running backwards through states 3, 2, 1 and then returning to 4), we essentially have the principle of a refrigerator. (We will take heat up at T_{low} and give heat up at T_{high}.) Of course, this needs power (electricity in most cases) to make such a cooling machine work.

The water in the steam engine can follow an entire cyclic path, though. First, it is heated in the boiler until it becomes steam, then the steam is cooled in the condenser where it is converted again into water, but now at much lower temperature. In principle, this low-temperature water can now feed back to the high-temperature boiler and the entire cycle can start again.

The moment supreme: entropy is invented

Yes, it's taken some time to prepare ourselves for the discussion on entropy. But let's not forget what we have achieved: we now know the precise meanings of absolute temperature, heat, energy, work, closed and open systems, and other important concepts. With these tools, we can now explore the concept of entropy. Because we do not want to use the massive mathematical arsenal which is used in typical scientific literature, we'll need to describe in words what entropy is all about. In the historical literature, you can find literally dozens of definitions of entropy. These descriptions are often referred to as the Second Law of Thermodynamics. Let's study a few and see how these definitions can help us understand the concept of entropy.

The originator, Clausius, formulated this as [Fast, 1962]:

2: The Science of Heat and Work

Heat can never, of itself, flow from a lower to a higher temperature.

Obvious as this statement seems, it is still a basic truth. Some might object that a refrigerator can "pump" heat from a cold interior to a warmer kitchen – but that does certainly not happen spontaneously since it takes a lot of electrical energy to do so.

Another important player, Lord Kelvin, formulates the concept as [Fast 1962]:

It is impossible to extract heat from a reservoir and convert it wholly into work without causing other changes in the universe.

Kelvin's definition is a very practical one. Imagine a ship equipped with an engine that can extract heat from the ocean waters in which it is moving. Since there is such a tremendous amount of heat available in the oceans, this would mean that we would have an unlimited amount of heat at our disposal for converting into work, Theoretically, we could use heat to traverse the oceans without the need for any other form of energy. The First Law would not forbid an engine like this – but the Second Law does, this heat-powered ship would constitute a perpetual motion machine of the second kind[36].

Now we're getting close to the supreme moment of understanding entropy. We have seen before that although ΔU is a thermodynamic state function, meaning dependent only on temperature but independent of the path from initial state to final state, ΔQ and ΔW are certainly not independent of which path has been followed. Both ΔQ and ΔW depend in general on an exact route for getting from A to B. Carnot, while studying his cycles, came to the conclusion that the parameter $\Delta Q/T$, however, is NOT dependent on the followed route. In fact, he was able to show going from V_1 to V_2, that $\Delta Q_{1,2}/T_{high}$ is equal to the corresponding quotient going from V_3 to V_4:

$$\Delta Q_{1,2}/T_{high} = \Delta Q_{3,4}/T_{low} \qquad (2.3)$$

---

[36] Second kind refers to the Second Law, the example is after Fast, 1962.

It was this conclusion which later led Clausius to introduce the concept of entropy, also called the Second Law of Thermodynamics:

$$\Delta S = \Delta Q/T \qquad (2.4)$$

Again, keep in mind that the entropy, as defined in equation (2.4), can only be determined for reversible process steps. Only in that case can the entropy be calculated from the change in heat.[37] However, for any process, whether reversible or irreversible, ΔS depends only on the initial and end conditions of the system, not on the path followed to get from one state to the other. In the next section we will show that for irreversible processes the entropy change is always larger than in the reversible case. This, as we will see, has important consequences for the efficiency of a given system to convert heat into work.

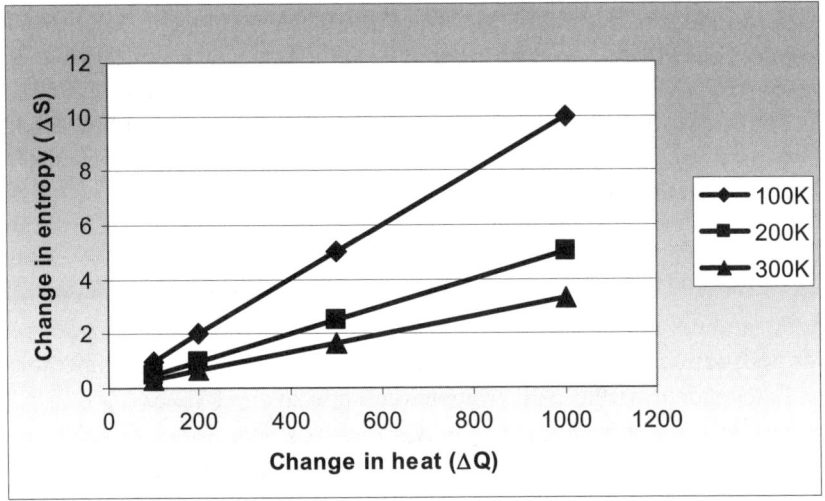

Figure 2.8 The entropy change with heat change at three different temperatures.

Figure 2.8 shows how the entropy changes with the change in heat at three different temperatures. It's important to note that, for equal

---

[37] Of course it is possible throughout to calculate the entropy for an irreversible step. Only we can then not use the simple equation $\Delta S = \Delta Q/T$ and we have to do more elaborate calculations which do need to concern us here.

2: The Science of Heat and Work

amounts of heat exchanged, the change in entropy is less at higher temperatures. Another way to look at this is depicted in Figure 2.9. We see that for a given amount of heat exchanged, the corresponding change in entropy is very large at very low temperatures. In contrast, at high temperatures we see that the change in entropy approaches zero. In Chapter 3 we will come back to these observations, and we'll show the fundamental background of this behavior.

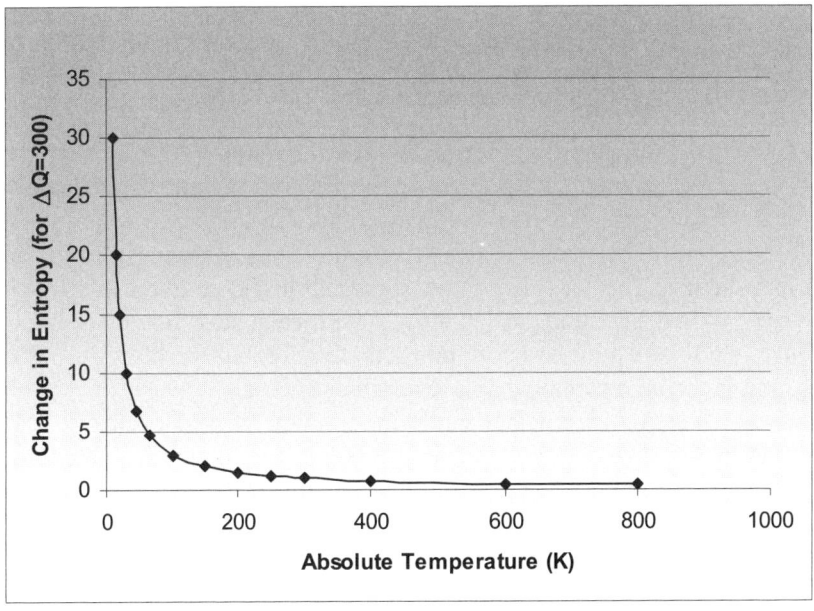

Figure 2.9 The accompanying change in entropy for a certain amount of heat (here put at 300) exchanged decreases as the temperature increases.

Entropy and reversible and irreversible processes

Another important point must be mentioned here. In the Carnot cycle described above, we took pains to perform the cycle under reversible conditions. However, if we had used irreversible conditions (for instance, by increasing the speed of the cycle), we would have consumed more heat from the hot reservoir to perform the same amount of work than we would have with a reversible process. As a result, our efficiency would have dropped! (We'll come back to this important point shortly.) Now, to get a better feel of the impact of reversible versus irreversible conditions let's look at an example. Most of us have used an air pump to inflate a bicycle or car tire. Today, we'll use a newer pump that has its own pressure gauge. Assume our tire is currently at 2 atmospheres but needs to be inflated to 2.5 atmospheres. We connect our air pump to the valve stem and start pumping. As we rapidly push down, we see the pressure dial move briefly up to about 4 atmospheres, and then go back to reveal the real tire pressure as we reverse the stroke. We realize the tire pressure was never 4 atmospheres, and so conclude that our rapid down stroke created a kind of a pressure shockwave, and that the pressure in the pump is not uniform. Also, we lose some of our work just in creating the shockwave and in overcoming the associated internal friction. Clearly we have created an irreversible process of inflating a tire.

If we wanted this process to be reversible, we would have to push the plunger down slowly so that pressure in the pump would go up just a tiny bit higher than the pressure in the stem, allowing it to release air into the tire. This way, the gauge shows almost exactly the pressure of the tire, and will continue rising slowly with each slow down stroke. Of course, we'll take a lot longer to pump up the tire, but we'll save on work – which is the big advantage of a reversible process.

We saw earlier that while the amount of heat exchanged and work done depends on the path one uses to go from the initial state to the final state, entropy – being a thermodynamic state function – is not affected by this choice. Thus, whether that path is reversible or not, the change in entropy (as with energy) depends only on the start and end conditions of the process. If we consider a cyclic process, as we did with the Carnot cycle, we must come to the conclusion that ΔS for the complete cycle must be zero, or to put it more precisely:

$$\Delta S_{1,2} + \Delta S_{2,3} + \Delta S_{3,4} + \Delta S_{4,1} = 0 \qquad (2.5)$$

But, remember, the transitions 2→3 and 4→1 were adiabatic, thus involved no heat exchange, and therefore $\Delta S_{2,3}$ and $\Delta S_{4,1}$ are zero. Therefore 2.5 can be reduced to $\Delta S_{1,2} + \Delta S_{3,4} = 0$. In the case where we have performed the Carnot cycle under reversible conditions, we can use $\Delta S = \Delta Q_{rev}/T$ to calculate the entropy from the transferred heat amounts and the expression above can be replaced by:

2: The Science of Heat and Work

$$\frac{\Delta Q_{1,2,rev}}{T_{high}} + \frac{\Delta Q_{3,4,rev}}{T_{low}} = 0 \qquad (2.6)$$

This equation states basically that although the heat $\Delta Q_{1,2,rev}$, coming from the hot reservoir or the heat source, is not equal to the heat $\Delta Q_{3,4,rev}$ (given up to the cold reservoir or the heat sink), the entropy increase in the expansion from V_1 to V_2 is the same as the entropy decrease in the volume contraction from V_3 to V_4.

Now, if the cycle (or part of it), had been carried out under irreversible conditions, equation (2.5) would still be 100% valid: the entropy change for the irreversible cycle is still zero but we *cannot* calculate the entropy from the heat changes anymore. In fact, under the irreversible conditions we will get: [38]

$$\frac{\Delta Q_{1,2,irrev}}{T_{high}} + \frac{\Delta Q_{3,4,irrev}}{T_{low}} \leq 0 \qquad (2.7)$$

This famous inequality was first noted by Clausius and is known as the Inequality of Clausius. The mathematical proof for the inequality is not difficult to fathom, but more important than the proof is that we understand the *message* from the Inequality of Clausius. That's because it has everything to do with the efficiency of the Carnot cycle.

Let's do an exercise to understand what's going on here. First, take a look at Table 2.2. In case A, we have imposed reversible conditions and therefore equation (2.6) holds. In cases B and C we have chosen the heat exchanges such that the inequality holds, and these are two examples of irreversible conditions. In case B, we take up 400 J from the hot heat source and dissipate 250 J into the low temperature heat reservoir, so that only 150 J is used to generate work. In case C, only 350 J of heat is taken from the high temperature heat source, and we allow 200 J to be given up again to the cold reservoir. We notice that both

---

[38] The symbol ≤ stands for equal or smaller, equal is true for reversible conditions and smaller is true for irreversible conditions. Because both sides of the expression are not equal, these kinds of expressions are called inequalities.

cases will satisfy equation (2.7). Cases B and C serve only as a qualitative illustration that irreversible conditions can come about in different ways, and that in reality a combination of B and C will typically occur. Important to notice is that the efficiency of the engine is always *less* in Cases B and C versus the reversible Case A! Thus, the efficiency calculated from formula (2.2) is a best-case scenario. In reality, the efficiency of engines will always be lower because of heat transfers following a non-reversible path, and because of other conditions, such as friction in moving parts.

Table 2.2 Efficiency of the Carnot cycle

Case	$\dfrac{\Delta Q_1}{T_{high}} + \dfrac{\Delta Q_2}{T_{low}} \leq 0$	Efficiency $\Delta W/\Delta Q_1$	Comment
A (Reversible)	$\dfrac{400}{400} + \dfrac{-200}{200} = 0$	50%	We take Q_1 as 400 J, T_1 at 400K, Q_2 as 200 J, and T_2 at 200K. In this example, we have converted 200 J into work.
B (Irreversible)	$\dfrac{400}{400} + \dfrac{-250}{200} \leq 0$	38%	Same heat uptake as in Case A, but we lose more heat to the cold reservoir.
C (Irreversible)	$\dfrac{350}{400} + \dfrac{-200}{200} \leq 0$	43%	Same as Case A, but now we take up less heat from the hot source.

To increase our insight into the difference between reversible and irreversible process steps, perhaps the following explanation will help. In general, we can say that for an *adiabatic* ($\Delta Q = 0$), *reversible* change in a gas volume, the change in energy is $\Delta U = -P\Delta V$ and that the pressure (P) can be expressed at any time as equal to the uniform pressure in the entire system (i.e, there are no pressure gradients). In case of an adiabatic process, *irreversible expansion* of a gas occurs as $\Delta U = -P_{ex}\Delta V$, where P_{ex} is the external pressure against which the expansion happens. So what does all this mean in terms of work? Well, suppose we have a volume of gas of 1 m^3 at 273K and 10 atmospheres. We decide to expand this volume to a final pressure of 1 atmosphere using three different processes, a) isothermal, b) adiabatic and reversible, and c) adiabatic and irreversible [example taken from Moore, 1972]. The results in terms of obtained final temperature, volume, and work done by this gas volume are listed in Table 2.3. We see that the irreversible adiabatic expansion has given us much less work (*about 50% less*) than the reversible expansion and thus provided much less work! This is because the gas has not cooled down as much as in the reversible case.

Table 2.3 Three ways to expand a gas at 273K and 10 atm and 1m^3 of volume to a final pressure of 1 atm.
(according to Moore, 1972, p 52)

Mode of expansion	End temperature (K)	End volume (m^3)	Work done (kJ)
Isothermal	273	10	-233
Adiabatic + rev.	109	3.98	-914
Adiabatic + irrev.	175	6.5	-547

We see now that reversible processes typically are very slow and that speeding up things causes processes to become irreversible. This is important to know, because it tells us that in real life situations where we want to do things fast, like drive a car or fly a plane, we have to pay a price in efficient processes. In Chapter 5 we will explore these observations and their consequences in more detail.

Entropy changes in spontaneous, isolated processes

We saw earlier that spontaneously occurring processes are always irreversible. The fundamental reason is that as soon as a process proceeds at a finite speed, it is almost always removed from equilibrium conditions (remember, equilibrium means that if we leave the system on its own, no further changes will occur). Now we'll consider a special case which is often seen in the real world, an isolated system. Isolated means that there will be no exchange of heat, work, or material with the surroundings and therefore, ΔQ and ΔW are zero. It can be easily shown that for an *irreversible, spontaneously* occurring process in an *isolated* system, the entropy always increases. The argument goes as follows: From equation (2.4) we know that for a *reversible* process, $\Delta S = \Delta Q_{rev}/T$. For an *irreversible* process, ΔS is always larger than $\Delta Q_{rev}/T$, or to put it mathematically:

$$\Delta S_{irrev} > \Delta Q_{rev}/T \qquad (2.8)$$

However, as we noticed above, for an isolated system there is no heat exchange with the environment, and thus $\Delta Q = 0$. When we combine this with expression 2.8, we come to $\Delta S_{irrev} > 0$. This is a very important conclusion: it states that *the entropy of a spontaneous process in an isolated system will always increase*! Many systems we encounter are isolated systems or can be made isolated. The best known is our universe. From a thermodynamic point of view, the universe is an isolated system and its entropy is continuously increasing, or as often stated, there is a continuous entropy production in the universe. This continuous entropy increase will eventually lead to a situation which is called the Heat Death. We will discuss this further in Chapter 6.

In Chapter 1, we learned that the mechanical theory of Newton could not explain why heat always flowed from hot to cold regions. Let's have a closer look now how the Second Law forbids the spontaneous flow of heat from cold to warm. Suppose we have two large containers filled with a liquid. One container is at 400K and the other at 200K. Imagine we transfer 200 J from the hot container to the cold one and assume that their temperatures will not change because of the large size of the containers. The total entropy change will be ΔS = -200J/400K + 200J/200K = +0.5J/K and is positive. Now let's try the opposite that we transfer 200 J from the cold reservoir to the hot reservoir. Again we can calculate the entropy change as: ΔS = 200J/400K −200J/200K = -0.5J/K, which is negative and indicates that there has been an entropy decrease! This would be in violation with the Second Law, which states that for a spontaneous process of an isolated system, the entropy change must be positive.

The First and Second Laws combined

We have seen that the First Law can be expressed as $\Delta U = \Delta Q + \Delta W$ (see formula [2.1]) and that the Second Law can be expressed as $\Delta S = \Delta Q/T$ (see formula [2.4]) or in a slightly different form also as $\Delta Q = T\Delta S$. If we now combine the two laws by substituting that the exchanged heat is equal to $T\Delta S$ into the First Law equation we arrive at $\Delta U = T\Delta S + \Delta W$. This can be rearranged into $\Delta U - T\Delta S = \Delta W$. We can draw two important conclusions from this. First, we can see that the amount of work we obtain from an amount of energy ΔU, is reduced by the amount of entropy production ($T\Delta S$). This is unfortunate and it will only become worse if we perform the process under *irreversible* conditions and we have seen earlier that the entropy increase for irreversible processes is larger than that of reversible processes (or $\Delta S_{irrev} \geq \Delta S_{rev}$). In that case we will get even *less* work out of the same amount of energy change! Second, when we keep the temperature constant, the term $\Delta U - T\Delta S$ can be rearranged into $\Delta(U-TS)$ and therefore:

$$\Delta(U-TS) = \Delta W \qquad (2.9)$$

What does this abracadabra bring us you will ask? The term $U-TS$ is called "free energy", also often called "available energy" in older literature. Now we see why we call it *available* energy since it is this amount of the total energy that is *available* to perform work.

(If you read the reference materials in the back of the book, you'll come across two more thermodynamic laws: the Zeroth Law and the Third Law. For the purpose of this book we do not need to concern ourselves with them, but if you're really curious, you can read about them in Appendix I.)

Perpetual motion and engines

Since the 11th century, many people have tried to beat the First Law with ingenious machines. There are good reasons for trying: if you could build a machine that could work forever without needing energy, that would solve the world's energy problem in one stroke. You would gain immeasurable wealth, fame, and surely the Nobel Prize as well.

Remember the tremendous excitement that emerged when Pons, et al. wrote about "cold fusion" in 1980? And that did not even involve perpetual energy production, but only a claim that nuclear fusion could proceed at room temperature instead of 5000°C – the temperature on the surface of the sun!

There are basically two different aims for perpetual devices: to achieve perpetual motion, and to generate work. A perpetual motion machine is typically not very useful other than its allure as a kind of magic show that can attract big crowds and so generate income from admission fees. In contrast, perpetual motion engines claim to generate work. (We call them *perpetual motion engines* because even perpetual motion requires that a certain amount of work be generated to overcome the friction forces, however small, that are present in all engines.) The hundreds of proposals made over the years for perpetual motion engines use many different forces to keep the movement going – including purely mechanical forces (gravity, expanding fluids, or springs) as well as magnetic, electrical, or buoyancy forces.

As said above, the First and Second Laws are postulates. This means no proof is possible, but it is merely based on many observations. Thus, in principle it could happen that tomorrow somebody builds an ingenious machine that can produce "free" energy. This would obviously be an enormous blow for the thermodynamic theory, but a blessing for humankind. Many such claims have been made, and many machines have been built either by people who intentionally produced frauds, or who were very serious about the matter and saw a mission to provide humanity with a useful tool. Also, many designs were made but never translated into real machines, and they are sometimes very difficult to prove wrong before they are actually built. Detailed mechanical analysis is often required before the flaw in the design can be found. The author knows no verified perpetual motion engine or machine has ever been built. Nevertheless, it's fun to look at some of these concepts.

Before we do, though, it's good to know that there are two kinds of perpetual motion engines, based on which of the two laws is being violated. Engines of the first kind typically claim that they can generate more energy (in the form of work) then the amount of energy that was put in, which clearly violates the principle of conservation of energy. Engines of the second kind are a bit trickier to describe. They try to convert heat into work without implementing any other change (achieving 100% efficiency), or purport to let heat flow from cold to warm, or attempt to convert heat into work without using two heat reservoirs at different temperatures. Simply put, second-kind perpetual motion engines draw energy from a heat reservoir and convert this heat into work without doing anything else.

Many perpetual motion engines of the first kind use the classic design of "overbalanced wheels." An early example comes from the Indian mathematician and astronomer Bhaskara, whose design incorporates tubes filled with mercury. In Figure 2.10, we see the operating principle of Bhaskara's idea. He claimed that the wheel would continue to rotate with great power, because the mercury in the tubes is not at the same distance from the axis at opposite sides of the wheel. Bhaskara probably never built a real device, but similar ideas later were incorporated into the designs of other inventors' engines, none of which ever worked. Fraudulent designs for perpetual-motion machines even made it to actual patents[39], which were later challenged in courtrooms. A famous example of a fraud was that made by Charles Redheffer in 1812 in Philadelphia. He claimed to have invented a work-generating perpetual motion engine, which seemed convincing until it was discovered that a man in an adjacent room was powering it.

Several famous names are connected to the idea of perpetual motion engines. Leonardo Da Vinci designed and built many devices and machines, including two devices to study the workings of perpetual motion. In his time, the principle of the conservation of energy was not known, but Leonardo had good insight into the working of machines and did not believe one could construct a perpetual engine. Simon Stevin, a Flemish scientist who lived from 1548 to 1620, actually showed that a purported perpetual engine based on a chain looped over a pair of asymmetric ramps would indeed not move without the addition of external energy.

An example of a perpetual motion engine of the second kind was provided by John Gamgee with his invention of the "Zeromotor" in 1880. His idea was to draw heat from the environment to let liquid ammonia boil; the ammonia vapor would expand and drive a piston. Afterward, the vapor was expected to cool down and condense, allowing

---

[39] Many patents can be found that claim to have invented a perpetual engine (for instance a patent for perpetual movement by Alexander Hirschberg in 1889, patent number GB 7421/1889). That these patents were granted was because in Great Britain patents filed before 1905 were not checked for whether the claims were realistic. This is unlike patents in the US where within a year a working prototype was required [van Dulken, 2000].

the process to start again. Gamgee proposed this idea to the American Navy as an alternative to its coal-fueled steamships[40]. The problem, however, was that ammonia at atmospheric pressure condenses only at temperatures lower than $-33°C$, and that temperature was not present in Gamgee's system. Thus, we see here a violation of the Second Law: if you want to draw work from heat, you must have two different heat reservoirs, one at a high temperature, and the other at a low temperature.

Interested readers may want to consult this book's references for more material on perpetual motion engines.

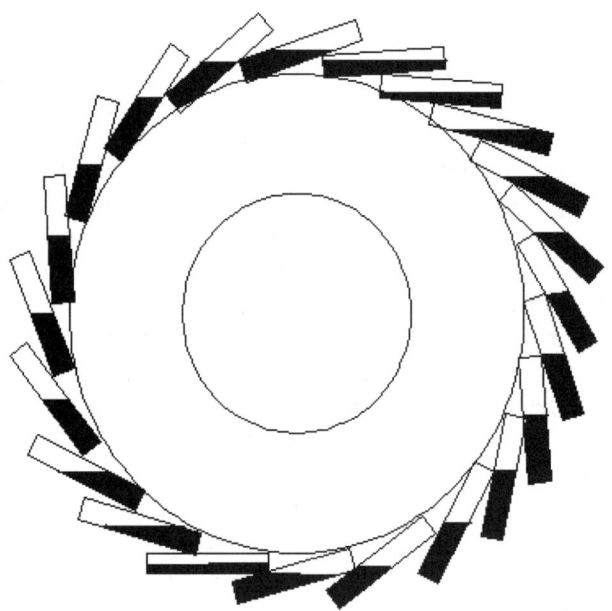

Figure 2.10 Perpetual engine after a design of Bhaskara.

---

[40] The American Navy was wrestling with the fact that their steamships were too limited in their routing because they could not get coal everywhere. Thus the Zeromotor was seen as a solution to this problem. The invention was even shown to President Garfield who was very positive about this approach.

Entropy and the direction of time

We saw earlier that the Second Law talks about a *continuous increase* in entropy for spontaneous processes. This is important, since it suggests that entropy is connected to the direction of time: a *continuous increase* can only be measured by at least *two* measurements at *different* times. An increase in entropy implies that the entropy measured later in time will be greater than the entropy measured at an earlier point. For that reason, the Second Law has been linked to the "forward" direction of time, which physics literature calls the "time arrow." The entropy concept proves that time can go in only one direction, because, using the Second Law, $\Delta S_{irrev} > 0$.

We can illustrate this with a simple example. Say we drop a sugar cube into a cup of tea. At first, the cube remains intact and visible, but slowly it falls apart and eventually disappears. (We'll learn later that the entropy of the cup of tea increases compared to the entropy of the sugar cube and the cup of tea separately). Since according to the Second Law entropy has increased, it must be that the cup of tea with the dissolved sugar cube occurs later than the situation in which the cup of tea and sugar cube are separate. If we filmed the dissolution of the sugar cube but showed the movie in reverse, anybody watching could immediately tell us that the film was being run in the wrong direction. The same principle holds for entropy production. If we measure the entropy of a system at two different times, we know that the measurement with the larger entropy will be made later in time than the measurement with the smaller entropy. In this way, entropy determines the direction of time. Another way to say this is that the First Law tells you only that at a T crossing, you can go either left or right as long as the principle of energy conservation is maintained, while the Second Law says you can go only left or right, depending on which direction results in an increase in entropy.

The linking of the Second Law to the time arrow has provoked a fierce discussion, not only among physicists but also among philosophers, on how to interpret the Second Law with respect to the development of time. There are two main camps, those who believe firmly in this interpretation of the Second Law and others who consider it nonsense. We'll discuss this topic in more detail in Chapter 3.

Let's take a break

Whew... that was difficult stuff, wasn't it? At this point, you might be thinking you won't make it to the end of the book. But relax, and remember that we've just covered in about 30 pages the thinking of many brilliant people over the past 150 years. And while we're appreciating that accomplishment, let's summarize our most important learnings so far:
- The desire to optimize steam engines led people to the insight that heat and work are different ways to transport energy from one region to another.
- The use of heat to generate work can never reach 100% efficiency under normal conditions. The efficiency is determined by the temperatures of the heat source and the heat sink in the engine. Irreversible conditions will always lower the efficiency of heat engines.
- Spontaneous, naturally occurring processes will always increase the entropy of an isolated system, or in other words, the entropy of the world increases steadily. This is also called the Second Law of Thermodynamics.
- Remember our hot tea cup in Chapter 1? The Second Law describes clearly that heat can never flow spontaneously from cold to hot.

There are a few other rules of thumb we can use to quickly assess what the entropy will do in a particular situation. Here are a few:
- In an isolated system, the entropy will always increase when spontaneous processes take place. In non-isolated systems (those where we can add energy), it is possible to let the entropy decrease in part of the system. However, this must always be compensated by increases elsewhere in some parts of the system or outside the system (the rest of the world).
- When a process makes temperature differences smaller, the entropy increases. This is due to the fact that smaller temperature differences (or gradients) imply that the energy is more evenly distributed in the system.

At this point, it may be good to say more about the origin of heat. Nature has a tendency to transform energy into heat. For instance,

2: The Science of Heat and Work

electrical energy converts easily into heat (look at the incandescent lamp, which has an efficiency of only a few percent); and friction always shows up in heated parts of vehicles (for instance, mechanical energy is turned into heat in the brakes of a car); and chemical energy often is released in the form of heat (as when burning wood). To understand why, we need to look at the nature of heat. Heat is produced by the movements of atoms and molecules, and these can move in several ways -- linear velocity, vibration, and rotation. Temperature determines the intensity of these atomic and molecular movements; the higher the temperature, the greater the velocity and other types of movements. Thus, matter can quite easily accommodate an increase in energy by raising the temperature. That is why heat plays such a predominant role in the transfer of energy from one location to another. Notably, heat's connection to the movement of atoms and molecules was discovered well after the First and Second Laws were formulated.

Now that our foundational work is done, we can explore how entropy has come to be applied to many areas of life. Starting around 1900, new mind-boggling insights about entropy began to emerge, and that is the subject of Chapter 3.

3
Much More About Entropy

Do we really understand what entropy is all about?

From the previous chapter, we have at least some idea of how the concept of entropy was introduced and how it can be used to understand certain phenomena, such as the direction of heat from hot to cold regions. We also have seen that while heat is transformed into work, a kind of waste occurs as some heat gets trapped into a form of energy that cannot be used to generate work, a phenomenon described by entropy. But why is this so? What is the fundamental reason for this naturally occurring inefficiency?

When I was a boy, magnets intrigued me. The headlights and taillights on our bicycles (I grew up in Holland!) were illuminated by small generators, each made from a coil of wire and a revolving magnet. Sometimes I was lucky and got my hands on a broken generator, which I disassembled to obtain the magnet. With the magnet you could do all kinds of neat stuff: make patterns in iron powder or pull the magnet over the ground in hopes of finding something valuable. I was very curious about how this seemingly magic attractive force could act over a certain distance without the magnet touching anything. Later on in high school, I learned about Maxwell's equations that described electromagnetic phenomena perfectly — only to be left with the perplexity you may have felt after reading Chapter 2: I still didn't understand how the invisible force was carried from the magnet to the iron powder. Yes, I could see its impact, but could not grasp how the force was transferred from the magnet to the powder. In similar fashion, entropy seems to defy understanding for many of us – and so it was with our predecessors.

A major leap toward a much better understanding of the fundamentals of entropy was made by an Austrian physicist, Ludwig Boltzmann,[41] who lived from 1844 till 1906. Boltzmann connected the classical definition of entropy to that of a new and emerging theory called statistical mechanics. This chapter will show how the statistical mechanical interpretation of entropy can substantially deepen our understanding of it.

Boltzmann's idea behind statistical mechanics was to describe the properties of matter from the mechanical properties of atoms or molecules. In doing so, he was finally able to derive the Second Law of Thermodynamics around 1890. But before getting into the details of his interpretation of the Second Law, let's have a look at the how atoms and molecules fought for a place in science.

History of the acceptance of the existence of the atoms in physics

Around 500 BC, Greek thinkers like Democritus were wrestling with the question of how many times you could divide a certain amount of material, such as a piece of stone. Could you divide the stone forever, or at some point would you encounter an ultimate quantity that would be undividable? ("Undividable" means that it is physically impossible to divide one more time, or that the remaining amount loses its characteristic properties if divided again.) The Greeks believed that there must exist a smallest quantity which would be undividable, and which they called an atom (from the Greek word *atomos*, meaning "indivisible"). Aristotle did not believe in atoms, thinking instead that

---

[45] Ludwig Boltzmann was born in 1844 in Vienna. He was a theoretical physicist who worked in various locations: Graz, Heidelberg, Berlin, Vienna. In 1902 he was teaching mathematical physics and philosophy in Vienna for which he became very famous. His statistical mechanical theory received a lot of criticism from his peers such as Wilhelm Ostwald. Because of these continuous attacks and his depressions he committed suicide in 1906 in Trieste (Italy). On his tomb one can find the famous formula $S = k\ log\ W$.

everything was composed of only four elements: fire, water, air, and earth. It took humanity more than 2000 years before the concept of atoms was reconsidered. That happened around 1800 when an English scientist, John Dalton[42], was studying the reactions of different gases such as oxygen and hydrogen, and nitrogen oxide and oxygen. As any high school chemistry student knows, hydrogen (H_2) and oxygen (O_2) react to form water (H_2O). But Dalton noticed that the two elements always reacted in the same ratio, with two parts hydrogen and one part oxygen completing the reaction. From this and other experiments, he concluded that hydrogen and oxygen must consist of atoms that reacted with each other to form water.

However, Dalton's theory did not distinguish between molecules and atoms. That distinction came later with the work of an Italian scientist, Amedeo Avogadro. Using Dalton's work as a foundation, Avogadro studied the formation of hydrochloric acid (the same stuff in your stomach) from hydrogen and chlorine. He noticed that the pressure resulting from the reaction was twice as large as expected, and concluded that both hydrogen and chlorine were made up of two atoms each. Thus, hydrogen gas consisted not simply of single atoms, but actual molecules (H_2)! The same was true for chlorine. Thus, Avogadro reasoned that one molecule of hydrogen reacted with one molecule of chlorine to form two molecules of hydrochloric acid. Chemists write this reaction as $H_2 + Cl_2$ → $2HCl$. With this work, science took the step from atoms to molecules.

Although more evidence later became available (for example, Michael Faraday's electrolysis work in 1830), scientists until the end of the century continued to engage in fierce debate over whether atoms really existed. The opponents of atomic theory were led by the influential Wilhelm Ostwald[43], who believed that atoms and molecules were not needed at all to explain the observations made at that time. As late as

---

[42] John Dalton (1766-1844) designed a system of chemical symbols, determined the relative weight of atoms, and showed that the consumption of chemicals involved in a reaction was always simple numerical ratio by weight. For this explanation he proposed the atomic view and can therefore be considered as the godfather of modern chemistry.
[43] Wilhelm Ostwald (1853-1932) is generally considered as the founder of classical physical chemistry. He published many outstanding books and publications in the fields of general chemistry, analytical chemistry, inorganic chemistry, and electrochemistry. In 1909 his work in catalysis and chemical equilibria was acknowledged by the Nobel Prize for Chemistry.

1891, Ostwald and Planck tried to convince Boltzmann that classical thermodynamics was sufficient to describe the experimental data of the day (Flamm, 1997).

It was in this climate that Boltzmann developed a connection between thermodynamic theory and atomic behavior, making his contribution only greater. After his death in 1906, the importance of Boltzmann's contributions was pointed out by Planck and Albert Einstein.

Statistical Thermodynamics: macroscopic and microscopic views

Before we can truly understand the results of Boltzmann's work, we should briefly look at some basic statistical tools. As soon as you try to derive conclusions from the consideration of individual atoms or molecules, you run into a problem: even tiny amounts of a material contain an enormous amount of particles. For instance, each gram of hydrogen gas contains roughly 10^{23} atoms. Clearly, it is impossible to follow all those individual hydrogen molecules, so scientists look at the average behavior of all atoms or molecules in a given system. By working with averages (average speed or direction of motion, for instance), they can predict the behavior of a system at macroscopic level (overall pressure, temperature, etc.).

It's all about probability

The following examples are mind-boggling and will explain how spontaneous processes tend toward an increase in entropy or as sometimes also said towards chaos. Imagine we want to construct a wall from black and white bricks. We want the wall to be four bricks wide and choose to alternate the different-colored bricks so that there are two black bricks and two white bricks in each row. How many layers can we make that are distinguishable to the eye? The answer can be found in Figure 3.1.

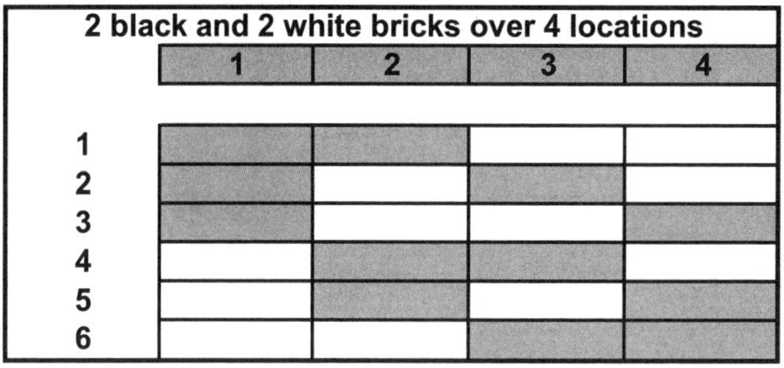

3: Much More About Entropy

	2 black and 2 white bricks over 4 locations			
	1	2	3	4
1	■	■		
2	■		■	
3	■			■
4		■	■	
5		■		■
6			■	■

Figure 3.1 Possible distribution of black and white bricks over four locations. Here we can distribute the bricks in six distinct different configurations.

We come to the conclusion that, despite all possible creativity, only six different layers are possible. Now, let's expand a little bit and assume that we want to use three white and three black bricks to construct a wall six bricks wide, and figure out by trial and error how many different configurations are possible. In Figure 3.2, all possible configurations have been laid out and the answer is that there are 20 different layers possible. We can continue with this process and figure out how many configurations would be possible with five white and five black bricks to build a 10-brick-wide wall. The answer is that 252 different layers of bricks are possible before we have to repeat the pattern.

Of course, it becomes more and more tedious to figure out how many distinguishable layers are possible. In Table 3.1, we have put down a few more possibilities. With 10 black and 10 white bricks, we have almost 200,000 different layers possible![44] It's shocking to see that with each increase in the number of bricks per layer, the number of different brick configurations increases sharply. For a 50-brick-wide wall, we have more than 100 trillion possibilities!

---

[44] It is clear that with this amount of possibilities, the method as used in Figures 3.1 and 3.2 is not practical anymore. As a matter of fact it is fairly easy to calculate the amount of possibilities.

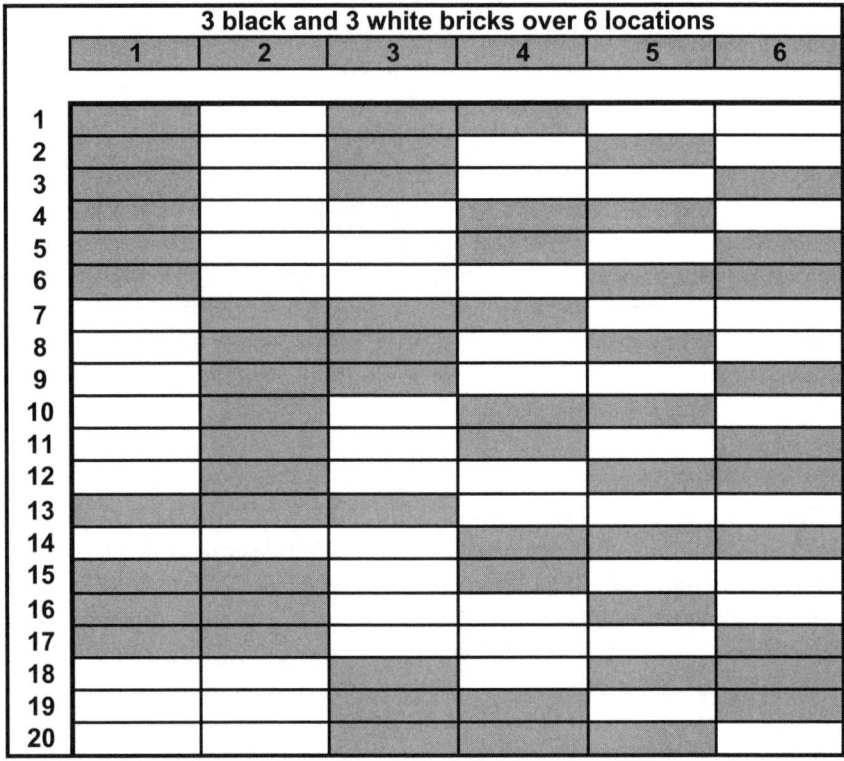

Figure 3.2 Possible distributions for black and white bricks over six locations. There are 20 distributions possible that are different.

Let's go back to the wall we were constructing in Figure 3.2 and have a look at the different layerings possible. What layers would we call chaotic, and which would we consider nicely ordered? Yes, I realize I'm asking for a subjective opinion along the lines of whether or not you like a certain painting; "chaos" does not have the same meaning for everybody. But I think most of us would consider rows 2 and 11, where both have alternate patterns of black and white bricks, as displaying a regular pattern. Layers like 4 and 5 would probably be seen as less nicely ordered. Suppose now that we help the contractor by handing him the bricks, but we give him black and white bricks at random, making sure only that we hand him an equal number of each color. After he's stacked many layers, we find that each of the possible layers from Figure 3.2 are present, but the chances of finding a particular layer (such as the orderly row 2) are only 1 out of 20, or 5%. Moreover, if we looked for the same orderly pattern in the 20-brick-wide wall, the odds of finding row 2 would be only in 1 in 184,756 or about 0.0005%. And in a wall 50 bricks wide, fewer than one out of 100 trillion layers would have that pattern! In everyday language, we would say that we probably never would find that particular layer (similar to our likelihood of winning the lottery, perhaps).

Table 3.1: Number of distinguishable layers as a function of width of the wall

Width of wall (in bricks)	Different brick layers possible
4	6
6	20
10	252
20	184756
50	1.26×10^{14}
100	1.01×10^{29}

Remember the example in Chapter 1 of the boy who dropped a can of marbles on the floor? Now we can better understand the fundamental reason why the marbles will form a chaotic pattern, and almost certainly will not align themselves in a chessboard pattern. The lesson here is that, while a chessboard pattern is not *impossible*, we would have to drop the can many trillions of times before we could create a checkerboard configuration. This is simply because there are so many more irregular patterns possible. Thus, the tendency of spontaneous processes toward chaos is purely a matter of probability.

Connecting entropy with atoms and molecules

In order to interpret entropy in atomic or molecular terms, we must introduce one more concept – that of macrostates and microstates. Let's start by having another look at Figure 3.2, where we constructed our wall six bricks wide and 20 bricks tall. If we rebuilt the wall by discarding the rule that each layer must consist of equal numbers of black and white bricks, and only required that we use 60 black bricks and 60 white bricks in total, then many more configurations become possible: for instance, we can have layers of entirely black bricks or wholly white bricks. The number of possible configurations for even such a small wall are dramatic: 9.66×10^{34}. Thus, this means that we can construct 9.66×10^{34} different walls using just 60 black and 60 white bricks! Each of these possible configurations is called a microstate.

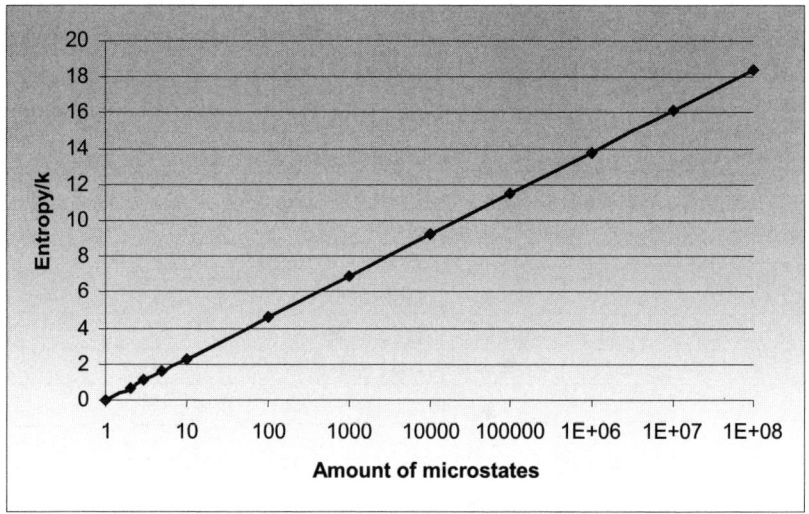

Figure 3.3 The figure S/k is dimensionless and can be plotted against the amount of possible microstates.

When there are microstates we also expect to find macrostates. A macrostate of a given system is characterized by parameters such as pressure, temperature, volume, and in our case of the wall, dimensions: width, height, and perhaps weight. Clearly, the dimensions or weight of the wall will not change with any of the 9.66×10^{34} microstates.

Now, let's give the number of microstates the symbol W, and thus W would have the value of 9.66×10^{34} in the example above. By carefully comparing the classical thermodynamic results obtained for the increase in entropy when two gases are mixed, or the entropy change when the volume of a gas changes, it becomes easy to link the classically

3: Much More About Entropy 59

defined entropy formula ($S = Q_{rev}/T$) to the number of microstates possible for a given system. The astonishingly simple result is contained in Boltzmann's famous formula:

$$S = k\, ln(W) \tag{3.1}$$

This means that the entropy, S, of a given system (say, a cylinder that contains air at a certain pressure, temperature, and volume) can be calculated once we have identified the number of microstates. In the above formula, the factor k is called the Boltzmann constant in honor of the inventor, and ln is the natural logarithm function.[45] In Figure 3.3 we can see how entropy evolves with the increase of microstates (W).

The physical meaning of formula (3.1) is that entropy will increase the more microstates are possible for a given macrostate. Perhaps this sounds a bit abstract, so let's look at an example. Consider the previously mentioned sugar cube dissolving in a cup of tea. We can use the Boltzmann formula to show how entropy will increase as the sugar disappears. Look at Figure 3.4. Imagine that the cube of sugar consists only of nine sugar molecules and that the tea itself is made up of 65 "volume elements," each of which can accommodate a single sugar molecule. At the start, when the sugar cube is still intact, the total entropy of the system is equal to the entropy of the sugar cube and that of the tea. However, since both the cube and the tea have only one microstate, the entropy of both systems is zero! When the sugar cube is totally dissolved, we have a different situation because now we can calculate[46] that there are about 30 billion microstates possible, which would lead to about 3.4×10^{-22} J/degree increase in entropy[47]. This is, of course, still a small number, because we have only a very limited number of sugar molecules and tea volume elements. For a real sugar cube and a

---

[45] $S=ln(W)$ means the value of $e^S = W$ where e has the value of approximately 2.33
[46] Here we follow the same approach as in the previous section and in Table 3.1. When we do that calculation then we find for the sugar cube case $3.2 \cdot 10^{10}$ possible configurations and thus microstates, which are about 30 billion microstates. A key assumption in Boltzmann's approach is that all microstates have the same probability of realization.
[47] By using Boltzmann's formula and the value of $k=1.4 \cdot 10^{-23}$ J/K we arrive easily at $S = (1.4 \cdot 10^{-23}) \times ln(3.2 \cdot 10^{10}) = 3.4 \cdot 10^{-22}$ J/K

cup of tea, the entropy increase according to Boltzmann's formula would be about 50 J per degree.

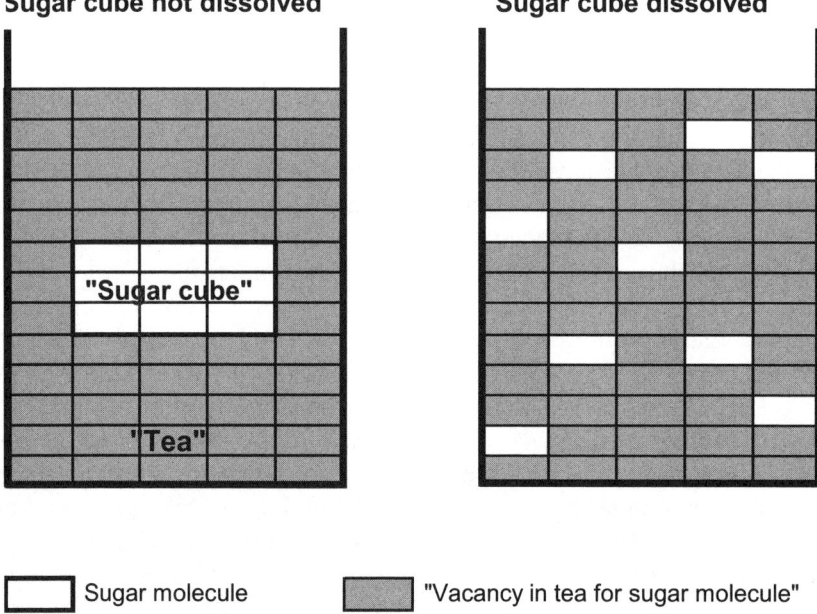

Figure 3.4 Very simple way to picture the dissolution of a sugar cube in a cup of tea.

The Second Law when the systems become real

Now things become very interesting. In the approach of Boltzmann, we deal with extremely large numbers and therefore use statistical techniques to draw conclusions about the averages of all the involved atoms and molecules. We have seen that entropy can be interpreted in terms of probability. The more microstates that are possible to realize a given macrostate, the more chance that we will observe the associated macrostate. It's the same with the state lottery: the jackpot has an entropy of zero because there is only one jackpot. Thus we realize that our individual chance of winning the jackpot is not very high, and the odds of winning it twice are even worse! It is not impossible, obviously – just very, very unlikely.

Figure 3.5 An example of micro machine made on silicon substrates. Typical dimensions are several micrometers or smaller. This is a six gear chain. All gears are driven sequentially by the drive gear (top center). Gear chains such as this one have been driven at speeds up to 250,000 RPM. (Courtesy of Sandia National Laboratories, SUMMiT^{TM} Technologies, www.mems.sandia.gov)

A nice, very early illustration that the impossible can sometimes happen can be found in the phenomenon of the Brownian movement, named after its discoverer, Robert Brown. Brown was a biologist studying the fertilization mechanism in flowering plants. One of his experiments in June 1827 was to suspend pollen grains in water. With his optical microscope, he then studied the suspended particles and observed that they were in restless motion. Of course, his discovery drew a lot of attention, since people linked the motion with the existence of life. It must be said, however, that Brown himself did quite a number of additional experiments which showed that the phenomenon was not linked to living materials exclusively, but occurred with inorganic particles as well. It took until 1905 before none other then Albert Einstein gave a good explanation. Einstein declared that the observed movement of the particles was caused by the continuous collisions from all directions of the particles with the individual water molecules. Normally, if the time interval is measured long enough, the net effect of these collisions will average out in terms of a net direction. The size of the pollen is thought to be about 2 μm and the number of collisions with the water molecules are several billions per second. Thus, although the collisions mostly cancel out each other, the balance is incomplete and so the particle moves a little bit, which is what Brown observed under his microscope. The Brownian movement can be considered the ultimate proof that in a gas or liquid, molecules are continuously in motion[48]. (It should be mentioned here that although the net effect of the collisions on each individual particle is greater than zero, when we look at all particles combined, no net effect is seen and all individual movements indeed cancel out.)

The important lesson from the Brownian movement is that even at surprisingly large dimensions (by atomic standards), we can already observe a deviation from our intuitive understanding -- that all collisions should cancel out, and no net movement should occur. So what does this have to do with entropy? Well, we're learning that the Second Law may not always be valid at small dimensions and tiny time intervals. Before we go into some of the consequences of this unsettling fact, let's first explain what's going on. Every time a particle obtains a net velocity in a certain direction, the entropy of the system is decreased because out of the random velocities (maximum chaos and maximum entropy) we have now a situation with less chaos and therefore a lower entropy. This, of course, is in violation with the Second Law, which states that entropy should increase for spontaneously occurring processes. We also can say that heat is extracted from the water and used to do work – namely,

---

[48] This is because of the kinetic energy each atom and molecule has and which is reflected in a velocity. Thus, although for the naked eye the solution appears to be in total rest, in reality there is a continuous movement in place. Because the direction of the velocity is completely random many collisions take place per time unit which in this case is shown by the movement of the pollen.

moving the particle. Since we are dealing with an isothermal system, this process should not be possible (remember, in

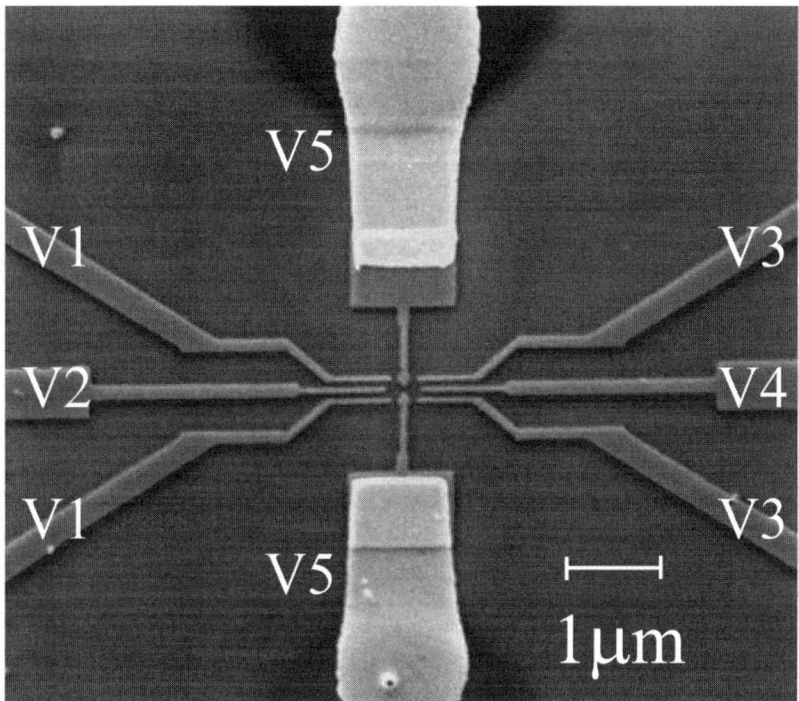

Figure 3.6 Quantum dots of about 50 electrons created in gallium arsenide which can be used to make logic gates (switches) for future generations of powerful computers (Courtesy Prof. Chang, Purdue University).

order to convert heat into work we must have a low temperature heat sink and a high temperature heat source at the same time in our system). This phenomenon can bring unexpected results when we go to the interesting world of microelectromechanical systems (MEMS). MEMS involves the integration of mechanical elements, sensors, actuators, and electronics on silicon through micro-fabrication technology – by the same technology that is used to make silicon computer chips. MEMS technology takes advantage of the fact that both the electronic and mechanical parts of the device use very similar fabrication techniques. The technology to manufacture MEMS devices (such as micro fluid pumps or collision sensors used in auto airbags) can fabricate in the nanometer range, well below the 2 microns we saw above for the dimensions of the pollen used by Brown. An example of a MEMS device is the six gear chain in Figure 3.5.

Because MEMS devices are so small, statistical averages may no longer apply and therefore unexpected behavior may occur. For instance, in case of a MEMS micromotor it is possible that the device deviates from the statistical description of the Second Law and runs for a short period while extracting heat from its surroundings. A similar complication may occur in microelectronics devices, where we are rapidly approaching the area of quantum devices (see Figure 3.6). Here only a few electrons will define the "on" and "off" condition of the transistor, and statistics of small quantities are quite different than the statistics of the large Boltzmann numbers discussed above.

Direct evidence that these deviations of the Second Law do indeed occur was done by following the trajectory of colloidal particles in water which were dragged by an optical tweezer [Wang, 2002]. Trajectories less than about 2 seconds in duration, showed indeed that entropy *decrease* was possible. However, over longer periods of time the particles followed again without exception the Second Law and an *increase* of entropy occurred.

Entropy and the direction of time: reprise

Time has been the subject of innumerable discussions, from the days of the ancient Greeks until today. Wouldn't it be fantastic if we could take a time machine and go back to the past? The fact is, time is a human invention. Several years ago, I read a series of articles in *The Humanist* about how people perceived time, which left me with several impressions. One is that time duration varies among people, and depends on many factors – a phenomenon called "subjective time paradox". For instance, time passes quickly when you're very busy, but waiting for something to happen seems to take forever. (As grandma said, "A watched pot never boils.") Also, the older you get, the faster time seems to pass – which seems to place older people out of sync with the quickly changing environment.

Then there is measurable time, which (as we will see) is also subjective. The mechanical clock was invented near the end of the medieval era; before then, people relied on sun clocks, which of course created differences all over the world in the duration of an hour. During the Industrial Revolution, the need for reliable train schedules required the synchronization of clocks in different cities; this was enabled by the invention of the telegraph. In 1885, an international conference agreed on a standard time (Greenwich Mean) and divided the world into 24 time zones. All this led to the perception that time was an immutable phenomenon, and not subject to change. The birth of the Theory of Relativity changed that perception considerably, which we'll discuss in detail in Chapter 4.

In Chapter 2, we touched briefly on the connection between entropy and time; now, we'll analyze that link more fully. We'll start with the atomic interpretation of Boltzmann's equation, which lets us take a much more fundamental look at the direction of time. Remember the dissolving sugar cube? We never expected the dissolved cube to reassemble itself spontaneously. Statistically speaking, such an event is not impossible, but it's extremely unlikely. This is because the cup in Figure 3.4 with unsugared tea and the cube has only one microstate (at least in our simplified representation), and the fewer the microstates, the less likely the macrosystem will occur.

Let's have a closer look at the situations sketched in Figure 3.7. In situation A, we start with a chamber divided into two compartments separated by a closed door. In the left compartment, a certain number of gas atoms are trapped. Because of their thermal kinetic energy, these atoms are in continuous motion and are colliding with each other and the wall all the time; arrows represent the magnitude and direction of their velocity. If we now open the door, we can expect the continuously moving atoms to pass through the opening and start occupying the right-hand compartment (thereby creating more possible microstates!). After some time, we will have approximately the same number of atoms in each compartment, although there will still be massive exchanges of atoms from left to right, and from right to left (the sketch in situation B reflects this). Because the situation is very dynamic, *in principle* we could expect that situation C would occur, and that all the atoms would return to the left compartment. But as we learned from previous discussion (and know intuitively), it is very, very unlikely.

Here we find the real reason why time flows only in one direction, and likely cannot be reversed. Entropy increases with more microstates (that is, it goes from situation A to situation B), and thus increases in the direction of the time arrow. For time to go backward, complexity would have to transform itself into simplicity once again. Thus, time is unidirectional, which is the same as saying that spontaneous processes are irreversible. Entropy is continuously increasing for spontaneously occurring processes as time goes on.

At the microstate level, we can see at once that there is no reason for irreversibility; the atoms can take any direction they want. At the atomic and molecular level, all processes can and will proceed in all directions. We cannot see a time arrow at this level, but it makes sense to attribute a direction of time to the changing ensembles of many atoms and molecules. As straightforward as this notion seems, there actually is a fierce debate among scientists and philosophers over whether entropy is connected with time at all. We won't go over all their arguments here, but appropriate references are listed in the back of this book.

Although a statistical treatment of Boltzmann provides much deeper insight into the nature of entropy and the origin of irreversibility, many scientists criticized his approach. Their complaints boiled down to Boltzmann's use of reversible mechanical laws of the movement of atoms or molecules in a gas to derive irreversible behavior. Scientists such as Loschmidt, Zermelo, and later Poincaré critiqued Boltzmann on this matter. Boltzmann, as usual, responded to these criticisms in a polite but very elaborate way – as we'll see below.

Point of zero entropy and of zero absolute temperature[49]

The Boltzmann's equation for entropy is, in fact, an expression for the absolute entropy, in contrast with the classical definition that says we can speak only about a difference or change in entropy between one situation and another[50]. The statistical expression for entropy shows that when a

---

[49] This paragraph discusses the relation between the zero point of the absolute temperature and the zero point of entropy. It is more of a theoretical nature and is mentioned here because it is important, however, not essential for what we want to achieve in this book. The phrase absolute temperature and absolute entropy refers to the situation that the temperature and entropy are measured with reference to a zero value or, to put it in other words, they are called "absolute".

[50] We have not mentioned this in Chapter 2, but classical thermodynamics defines the entropy only in terms of changes between one state and another state and has no clear zero value for entropy defined.

given macrosystem has only one microsystem available, then $W = 1$ and thus $S = k.ln(1) = 0$. Intuition suggests that there will be a connection between the lowest temperature possible (absolute zero) and the value of

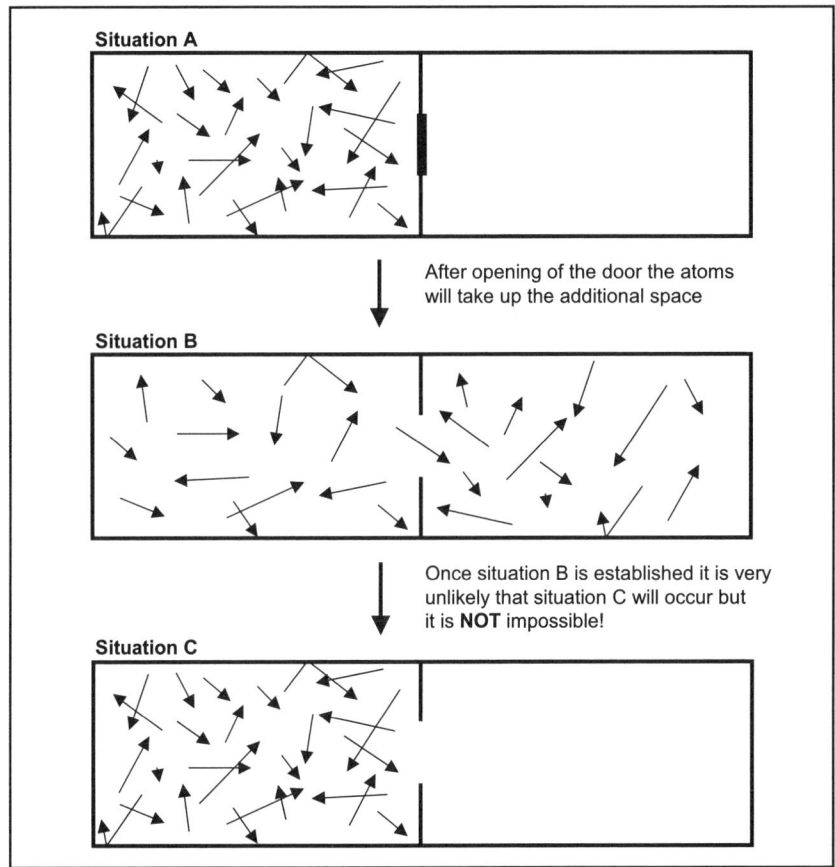

Figure 3.7 Sketch of a spontaneous, irreversible process. The gas atoms enclosed in the left compartment in situation A will, upon opening of the door, spontaneously move into the entire available space (situation B). It is extremely unlikely that from situation B the atoms all will move back to the left compartment and that situation C will occur.

the entropy being zero. This connection is described by what is often referred to as the Third Law of Thermodynamics: Nernst's heat theorem[51]. Nernst's heat theorem states that at the absolute temperature of zero Kelvin, all atoms will occupy the lowest energy level available. When that is the case, all processes become adiabatic because heat is no longer being exchanged. This means that for processes occurring at absolute zero, the change in entropy will be zero (recall $\Delta S = \Delta Q_{rev}/T$, and for adiabatic processes $\Delta Q_{rev} = 0$). Max Planck extended Nernst's principle and postulated that there is only one energy level available (or perhaps better occupied) and therefore the entropy will be zero, since only one microstate is available. This situation is best appreciated in a perfect crystal of an element at absolute zero Kelvin, since in this simple case there will be only one possible microstate available[52].

Boltzmann's struggle with the scientific community

As pointed out previously, Boltzmann experienced considerable opposition to his ideas from formidable opponents such as Nernst, Mache, and Ostwald. This was directly due to the fact that the acceptance of atomic theory was not a done deal. At that time, nobody had ever "seen" atoms, and they were considered merely a theoretical aid to help explain certain phenomena. However, one would assume that in the present day, where we can really see atoms using high-resolution electron microscopes (see Figure 3.8), Boltzmann would get the credit he deserves. But guess what: almost 100 years after his death, scientists are still arguing over how to interpret Boltzmann's ideas on entropy. Joel Lebowitz has given a good overview of objections and interpretations in the September 1993 issue of *Physics Today*. Lebowitz uses a nonsense approach in defending Boltzmann's extraordinary achievements against all kinds of scientific objections. We won't go into all the details, but it's

---

[51] Walther Hermann Nernst (1864-1941) was a brilliant physical chemist. He was a student of Wilhelm Ostwald and received the Nobel Prize in 1920 for his work in thermochemistry. He is best known to chemists for his "Nernst equation" that describes the voltage of an electrochemical cell as a function of the ionic concentrations of the electrolytes. His thermochemical work led to the formulation of what is often called, the third law of thermodynamics (see text for more details). It were Planck and Nernst, two of the most recognized scientists of that time, who convinced Albert Einstein to come to Berlin in 1918.
[52] There can be exceptions to this rule (degeneration of energy levels) but this would go beyond the scope of this book.

interesting that in the subsequent issue of the journal, there appeared five reactions from subject-matter experts who supported or contradicted Lebowitz' bold statements. This shows that the entropy debate is far from over, both from the theoretical point of view and even in the practical world of nanoelectronics and MEMS.

Figure 3.8 Transmission Electron Microscope (TEM) image of an atomic lattice. Here, a layer of amorphous HfSiO is trapped between layers of crystalline silicon (bottom) and metallic tantalum nitride (dark top layer). The contours of the individual atoms in the mono-crystalline silicon lattice are visible (Courtesy Brendan Foran, SEMATECH).

Energy efficiency and some conclusions

At the beginning of this chapter, we were looking for the fundamental reasons behind ever-increasing entropy and the finding in Chapter 2 that in the steam engine not all energy can be transformed into work and efficiencies were always substantially smaller than 100%. Now we can understand this better. Before we start up a steam engine to generate work, we have an amount of coal and a bucket of cold water. We fill the boiler of the steam engine with the cold water and ignite the coal in the furnace to heat the water. Steam will form and drive the pistons of the steam engine. In addition to the steam formation we also generate a lot of hot gases from burning the coal. After some time all the coal is burned and the steam engine cools down. In the condenser we have collected cold water from the condensation of the steam. Thus at completion of the process we are left with ash, a bucket of cold water and, of course, a certain amount of work we have generated by our steam engine. However, we also created entropy because the entropy of hot gases and ash is higher than the entropy of coal. We can also say that coal is a more ordered substance than the hot gases and ash it is converted in. From Chapter 2 we know that the First Law demands that the amount of energy (ΔU) enclosed in the coal is converted into an amount to generate work (ΔW) and an amount that is used to produce entropy (ΔS) (recall $\Delta U = T\Delta S + \Delta W$). It is the factor $T\Delta S$ that is responsible for keeping the efficiency smaller than 100%[53].

We found that Boltzmann's atomistic interpretation of entropy gives us greater insight into the secrets of how nature really works. It shows that ever-increasing entropy is simply nature's progress toward more probable configurations. Another way of stating this is to say that nature tends to go from order to disorder. This progression also explains why spontaneously occurring processes are virtually irreversible, and why the arrow of time is *almost* certain to go in only one direction.

---

[53] At the risk of confusing the reader I refer to the discussion on fuel cells in Appendix VIII. In fuel cells the chemical energy enclosed in fuels is directly converted into electrical energy. Thus no heat conversion is involved and as a result the efficiency can be much higher than in coal fired power plants.

4

Link of Thermodynamics to Modern Physics

Three men and thermodynamics

Now, we come to the discussion on modern physics. In the decade from 1895 to 1905, three very important discoveries were made: J.J. Thomson proved the existence of the electron, Becquerel identified natural radioactivity, and Rontgen discovered X-rays. In addition, two important theoretical developments were put in place: Max Planck's hypothesis that radiant energy in its interaction with matter behaves like particles with discrete quanta of energy, and Einstein's Special Theory of Relativity. These events defined the beginning of modern physics (Semat, 1939)[54]. One can argue that the birth of Quantum Mechanical Theory followed by the Theory of Relativity has absolutely changed the way physicists (and many other professionals) regard the world. In this chapter, we will discover how Boltzmann (whose new concepts included atomistic views and events based on probability instead of certainty) drove Max Planck to lay the foundation of Quantum Mechanical Theory. We will also see that thermodynamics inspired Einstein to develop the Theory of

---

[54] In this chapter, we'll use the term "classical physics" to describe physics before the advent of quantum mechanics and relativity theory.

Relativity[55]. In fact, many brilliant scientists who played a role in the development of these theories were very familiar with developments in thermodynamics. To illustrate this point, we'll describe the interactions with thermodynamics of Planck, Einstein, and Erwin Schrödinger. (Schrödinger was the inventor of wave mechanics, a method of describing atoms that became an essential part of quantum mechanical theory.) All three were Nobel Prize-winning physicists: Max Planck in 1918, Albert Einstein in 1922, and Erwin Schrödinger in 1933.[56] An interesting thermodynamic connection among the three is that they all worked for some period at Berlin University.[57] As we will see in this chapter, all three scientists were not only strong supporters of the concepts developed by Boltzmann, but also were experts in thermodynamic theory.

While we won't go into an extensive review of quantum mechanics and relativity, we will spend a little time reviewing the most important aspects of how these theories caused a true paradigm shift in physicists' view of how nature really works. These ideas remain as valid today as when they were first conceived (as does thermodynamics). We'll also look at whether these modern theories changed any aspects of the fundamental laws of thermodynamics.

Why couldn't Newton's mechanics explain everything?

Dark clouds for classical physics

Let's start by having a look at Ohm's law, which says that the magnitude of an electrical current through a wire is proportional with voltage. It also

---

[55] In 1905, the Special Theory of Relativity was published, followed by the General Theory of Relativity in 1916. We will touch only on the Special Theory in this book.

[56] Boltzmann should have received a Nobel Prize for his earth-shifting work, but unfortunately never did.

[57] In the period 1871 to 1931, the University of Berlin was the world's leading institute for thermodynamics and is therefore also called the Berlin School of Thermodynamics. All three fundamental thermodynamic principles, energy conservation by Hermann Helmholtz, entropy law by Rudolf Clausius, and the zero entropy at absolute zero temperature by Walter Nernst were established while their inventors were connected to this institute (of which Helmholtz and Clausius were the founders) [Ebeling and Hoffman, 1990].

4: Link of Thermodynamics to Modern Physics

says that the voltage across the wire is proportional to the magnitude of the current and the resistance of the wire. The simple formula is this: $V = I \times R$, where V is the voltage across the wire, I is the magnitude of the current, and R is the resistance of the wire. Ohm's law, although very practical in many electrical calculations, is a phenomenological description and from that point of view very similar to the First or Second Law of Thermodynamics. It describes quite accurately the voltage-current characteristics of an electrical circuit, but does not explain why the resistance of a copper wire is lower than that of an aluminum wire, or why glass is an insulator and transparent to light, while metals do conduct electrical current but are opaque.

There were other observations that classical theories could not explain, such as:

- The phenomenon of "black body" radiation that described the energy emitted by a radiant source as a function of temperature
- The photoelectric effect that showed the emission of electrons from a surface into a vacuum under influence of light
- The famous Michelson-Morley experiment, which attempted to measure how the velocity of light was affected by the movement of a hypothetical "ether" that was presumed to fill the cosmos, including outer space.

Because classical theories failed to explain the results of the Michelson-Morley experiment and the phenomenon of black body radiation, Lord Kelvin called them around 1900 "the two clouds on the horizon of physics," but no one at that time was overly concerned that this would shake the foundations of the classical framework[58]. However, these results presaged the end of Newtonian physics, triggering the births of Einstein's Special Theory of Relativity and Planck's quantum theory.

---

[58] In fact, when the young Max Planck had to decide which profession to enter (he was also a very talented musician), his teacher strongly discouraged him from studying physics because "everything was already discovered and understood, and no new things were left to be understood".

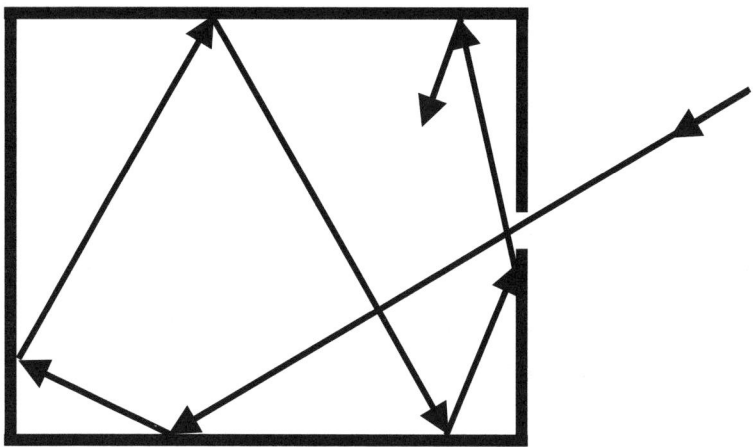

Figure 4.1 When light enters a cavity through a small hole it can almost not escape to the outside world. The hole itself approaches therefore the characteristics of what is called a "black body".

Black body radiation

All surfaces above absolute zero emit radiation[59]. At low temperatures (room temperature to a few hundred degrees), heated objects such as an iron rod show no visible radiation – although we can feel the invisible, infrared radiation as heat. However, when the rod is heated to about 600°C, it starts to glow a dark red color. Heat it some more, to about 800°C, and the rod radiates a lot of yellow light. For ages, blacksmiths have used the color of iron to estimate its temperature, and decide whether it's hot enough to start beating into a new shape. This principle also is used in the everyday light bulb, whose tungsten filament is heated to around 3000°C to generate a very bright, whitish light.

Although we've known for centuries about heated metal's ability to produce light, classical physics, as we will see below, could not explain this phenomenon. And just as it took Chapters 2 and 3 to explain the flow of heat from hot to cold, we now face a similar situation in

---

[59]Radiation by electromagnetic waves. Electromagnetic radiation is characterized by its wavelength and its intensity.

4: Link of Thermodynamics to Modern Physics

trying to understand why an object shifts color as its temperature increases. Around 1900, a lot of experimental work was done on the determination of the wavelength and intensity of the radiation of a "black body" at different temperatures. A black body is a surface that absorbs all electromagnetic radiation[60]. Although there are no materials that fully meet this definition, a good approximation of a black body can be made by constructing a box with a small hole in its side (see Figure 4.1). Nearly all of the light entering the pinhole will remain in the box, no matter what its interior color or material[61]. The box then will emit radiation, which will increase as the interior (or cavity) of the box grows warmer.

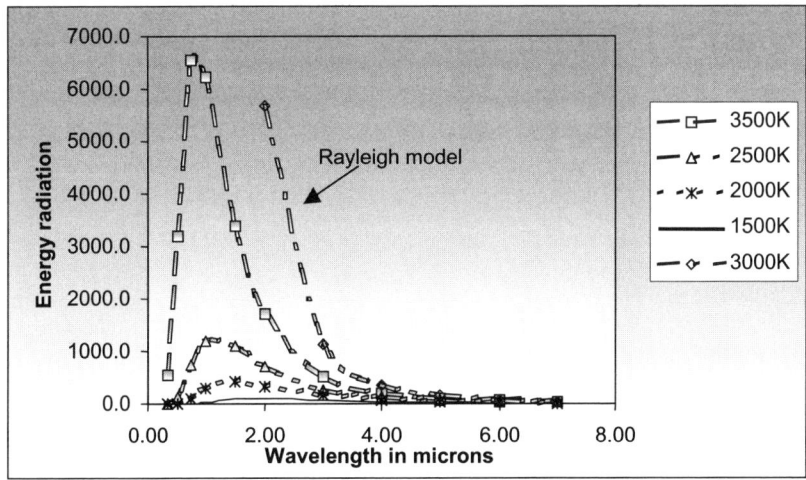

Figure 4.2 The radiation intensity as a function of the wavelength of a black body for different temperatures according to Planck's theory of light quanta. For 3000K the intensity radiation distribution as calculated from the classical Raleigh model is included as well.

Black body radiation intensity can be measured as a function of the wavelength, as shown in Figure 4.2. We see the maximum intensity

---

[60] Thus, it also absorbs all visible light and will therefore appear to be totally black.
[61] Our eyes can be considered as black bodies. Although light can enter the eye through the pupil, it cannot escape. That's why pupils are always black!

increasing with temperature (which we would intuitively expect) and shifting toward shorter wavelengths at increasing temperatures. However, the most important deduction from Figure 4.2 is that there appears to be a maximum in the profile: at large and short wavelengths, we see hardly any radiation intensity at all. This observation was in severe conflict with classical theory. In the classical model, the radiation of the black body cavity was thought to be generated from standing waves within the cavity. The shorter the wavelength, the more wave modes (i.e., ways that a wave can fit within the cavity) are possible. Rayleigh and Jeans carefully evaluated the associated radiation of all the different modes, and the result was that the intensity of the radiation would sharply and continuously increase at shorter wavelengths while predicting no maximum. This prediction was also known as "the ultraviolet catastrophe." Although some other attempts were made, no acceptable explanation could be found. Then in 1900, Max Planck developed an explanation that changed the scientific world. He assumed that oscillating electrons in the surface could only give up energy in discrete packages, which he called "quanta." The energy content of a quantum was proposed to be equal to

$$E = h\nu \quad (4.1)$$

where h is the constant of Planck and the symbol ν stands for the frequency.[62] Thus with increasing frequency (or shorter wavelength), the energy content of the quanta increases. Because the total amount of energy available for radiation is limited (there cannot be more emissions than we inject in the form of heat!), there will be fewer high-energy quanta available to form the wave, and consequently we see a drop in radiation at the shorter wavelengths.

---

[62] We recall the relationship between frequency and wavelength: $c = \lambda \nu$ where c is the velocity of light and λ is the wavelength. From this, one can write $\nu = c/\lambda$ and thus Planck's equation also can be written as $E = hc/\lambda$.

4: Link of Thermodynamics to Modern Physics 77

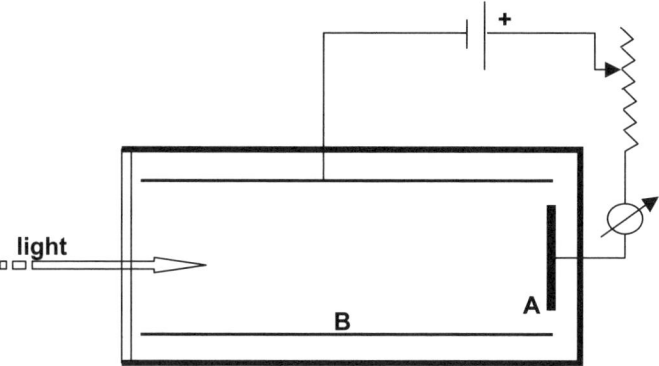

Figure 4.3 The experimental system that Philippe Lenard used to study the effect of light on the emission of electrons from a surface in vacuum.

The photoelectric effect

More bad news was underway for classical physics. In 1887, Hertz found that a spark from an electrode would occur more readily if the light from another spark illuminated the electrode. During the subsequent 15 years, these experiments were refined. In 1902, Philippe Lenard (Nobel Prize winner in 1905) studied this phenomenon, using the vacuum tube apparatus shown schematically in Figure 4.3. In a vacuum tube, there are two electrodes, A and B. Between the electrodes is a voltage difference that can be varied so that B is negatively charged and will repel electrons. Thus for a given negative voltage across the two electrodes, the electrons must leave A with a minimum of speed (energy) or they will not make it to B. Through the glass wall of the tube we can shine light directly on A and with the current meter measure how much current runs between the electrodes. The idea was that the light would liberate electrons from the surface with enough kinetic energy (or speed) to reach B. While traveling from A to B, they would be slowed down by the negative voltage from electrode B and would give up their kinetic energy (or speed). Lenard made two important observations: there was a certain voltage (V_0) where all electrons were repelled from electrode B, and no current was observed. This means that the energy of the electrons that are released from the surface of electrode A has a maximum value. The most important observation was that V_0 depended on the *frequency* and not the *intensity* of the light, since from a classical point of view the amount of energy was associated with intensity. This was a very puzzling situation for physicists at that time since, assuming light was a wave, the expectation was that V_0 would depend only on the intensity.

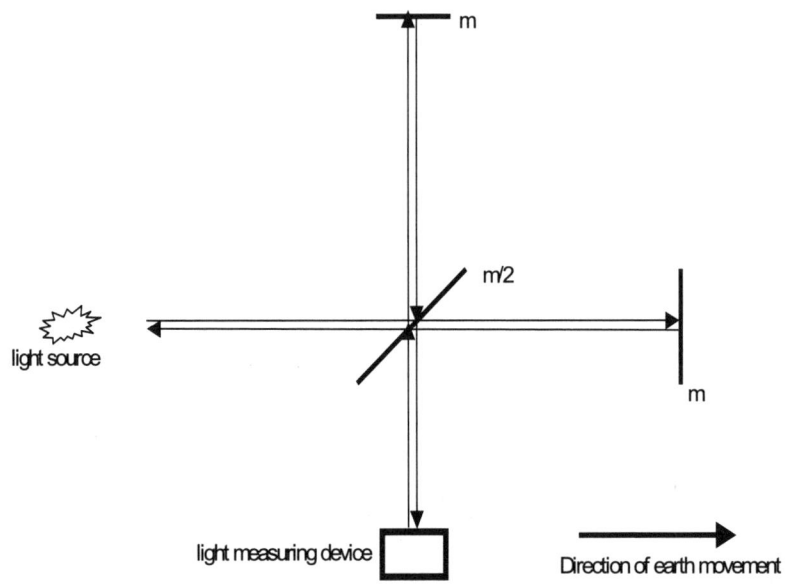

Figure 4.4 The experimental setup as used by Michelson and Morley in 1887 to prove the existence of an ether. This experiment is one of the most famous in scientific history.

In 1905, Einstein published his famous paper, "Über einen die Erzeugung und Verwandlung des Lichtes betreffenden heuristischen Gesichtspunkt," *(On a Heuristic Point of View Concerning the Production and Transformation of Light)* which appeared in the prestigious *Annalen der Physik*.[63] *(Annals of Physics)* In this work, Einstein showed that the observations of Lenard could be explained by assuming that light could be considered as mass less particles assembled into discrete energy packets of size $h\nu$, exactly as the idea Planck had expressed in his black body radiation theory. The light packets were later named "photons." With this explanation, it became clear that below a certain frequency of light, there was no current, since the energy of the

---

[63] This work won Einstein the Nobel Prize for Physics in 1922.

photons was simply not enough to liberate electrons from the surface, or to pass over the voltage barrier between the two electrodes.

The Michelson-Morley Experiment

Around 1880, the scientific world was still wrestling with the question of how electromagnetic waves (light) could propagate in vacuum. This, of course, was driven by the knowledge that water waves need water to travel, and sound needs air to propagate (remember the high school experiment of the bell in the vacuum jar). Likewise, it was assumed that there also must be a medium to let light travel in vacuum. This substance, which was assumed to be invisible and with no mass, was called "ether."

In 1887, two scientists designed an experiment to show the existence of this mysterious ether by a very clever experiment. Their idea was to project a split light beam in two orthogonal directions. The light beam was split with a semi-transparent mirror (m/2 in Figure 4.4). At the end of each light path was a mirror (m) to reflect the beam back. The apparatus was positioned so that one of the two optical paths could be aligned with the direction of the earth's movement. The ether was assumed to have the same velocity as the earth, so that the aligned beam would have the velocity of light plus that of the ether on the way out, and the velocity of the light beam minus that of the ether on the way back[64]. In contrast, the orthogonal beam (vertical beam in the figure) would not be affected by the velocity of the ether, and would act as a control beam. Because of the difference in velocity of the two directions, it was expected that an interference pattern would arise at the light-measuring device because of the difference in phase of the two light beams (that is, the phase difference would show up as an increase or decrease in light intensity, a phenomenon called "interference"). However, whatever variations Michelson and Morley tried (by rotating their apparatus, for example), they were not able to generate an interference pattern.

This outcome puzzled the scientific community tremendously, since it was so much in conflict with Newtonian mechanical theory. According to Newton, the velocities of the ether and the light should combine to produce the final speed of the light beam. But this famous

---

[64] This assumption was analogous to the everyday experience of having a tailwind at your back when riding a bike outward, and a headwind when coming home. Obviously, your speed will be faster when going out than when returning.

experiment showed otherwise: regardless of the speed of the assumed substance (i.e., ether), the speed of the light beam remained constant at about 300,000 kilometers per second.

The newly revealed fact that the speed of light remained the same regardless of the speed of the observer was the most important motivation for Einstein to develop his Special Theory of Relativity. In that theory, Einstein did not really attempt to explain the behavior of light, but merely started with the postulate that the velocity of light is invariable and does not change with the velocities of whatever system it is in. Investigating the consequences of this assumption, Einstein came to the dramatic conclusion that time and distance are not constant, but vary according to the velocity of the system in which they reside. We'll talk more about this concept later in the chapter.

The connection between the classical mechanics of Newton, the Quantum Mechanical Theory, and the Special Theory of Relativity

How do all these theories fit together? Fortunately, the answer is quite simple. The mechanics of Newton is still very much valid. We still use it to calculate Space Shuttle orbits or how long it takes to stop a train. Newton's formulas, however, start to break down in the very small world where atoms and electrons live, and whenever velocities start to approach the speed of light. So we can consider classical mechanics as a special case of both quantum mechanics and relativity theory. If a given speed is much slower than light, Einstein's formulas transform into Newton's. The same is true for quantum mechanics: when objects are much larger than atoms, Newton's formulas again hold true. In Figure 4.5 a more graphical outline is pictured that maps out the application field of the different physical theories.

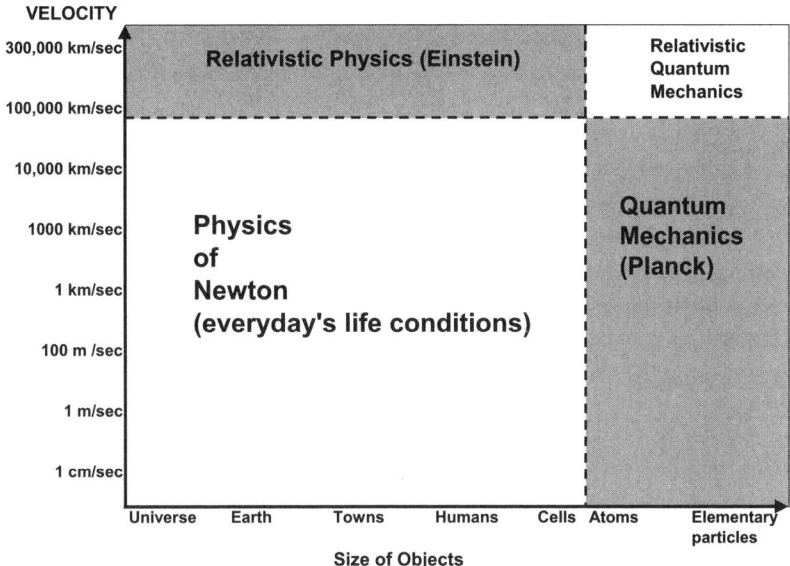

Figure 4.5 Mapping of the field of application of the laws of Newton, Einstein, and Planck.

Thermodynamics at the birth of modern physics

Boltzmann's heritage

In Chapter 3, we saw how Boltzmann's work changed people's insight into entropy. Finally, we could explain why heat flowed from hot to cold, rather than just using a postulate to describe that situation. In retrospect, it's clear that Boltzmann's work strongly influenced the dramatic developments in physics around 1900. (Flamm, 1997). These developments are, of course, the Quantum Mechanical Theory (starting in 1901) and the Special Theory of Relativity (arriving in 1905). As we will see in more detail below, two important concepts used in quantum mechanics are situations occurring through probability rather than with the absolute certainty associated with classical mechanics, and the existence of discrete levels of energy rather than an energy continuum. The introduction of the statistical mechanics (and thus probability) in the interpretation of the Second Law of Thermodynamics, (therefore delivering another push to the acceptance of atomistic concepts), paved the way for Planck in explaining black body radiation. The mathematical connection between entropy and the probability that a certain macrostate will occur was derived by Boltzmann in a very extensive (84-page) article in 1872, using an impressive mathematical treatment. (Boltzmann was a very good mathematician who taught mathematics at the University of Vienna from 1873 until 1876). In this article, he introduced his famous H-theorem that allowed him to make the connection between entropy and the amount of microstates it takes to make a macrostate. In this work, he also introduced discrete energy levels needed to obtain mathematical solutions for complicated equations. This principle later was used by Planck to lay the foundations of quantum mechanics.

Subsequently, in a course on natural philosophy in 1903, Boltzmann pointed out that space coordinates and time should be treated equally (Flamm, 1997), which is exactly what Einstein did in his Special Theory of Relativity. As mentioned earlier, both Planck and Einstein acknowledged Boltzmann, after his tragic suicide at age 62, for his great contributions to modern physical theories.

What Planck thought about thermodynamics

Around 1900, Planck was an expert in classical thermodynamics and wrote many articles and books about that theory. The concept of entropy especially held his interest, but he published also in the fields of dilute solutions and thermoelectricity. Of course, being a time-oriented fellow, he was familiar with the results of Boltzmann's works. However, being a physicist of the "old school," he was raised without having the concept of atoms in his scientific toolkit. In 1891, for instance, he and Ostwald had a discussion with Boltzmann at a conference where Planck stated that thermodynamic methods without the incorporation of atomistic models were sufficient to explain those days' physical observations. Also, Planck was not very pleased with the statistical approach of Boltzmann [Flamm, 2000]. His main objection was that Boltzmann's statistical approach allowed that the change of entropy for spontaneous processes could become negative (i.e., an entropy *decrease*), although at an extremely low probability (see Chapter 3 for more details on this topic).

But Planck was wrestling at the turn of the century with understanding black body radiation behavior. Since 1861, when

4: Link of Thermodynamics to Modern Physics 83

Kirchhoff[65] first described a black body, the radiation behavior was studied and described by a slew of well-known physicists such as Wien, Stefan, and Boltzmann. However, all these attempts led only to a radiation law that had very limited applicability. The breakthrough in Planck's understanding came when he started to use Boltzmann's statistical approach. In fact, it was Planck who wrote the current well-known form of the Boltzmann equation, $S = k\, lnW$, in his famous 1901 article [Planck, 1901 *Annalen der Physik*]. It was in this text that Planck proposed that the radiation might consist of small packets (*quanten*) of size hv. This was the beginning of quantum mechanical theory. Planck struggled a long time with his own thoughts, since they were so in contrast with the classical belief of continuous energy. For some time he saw his *quanten* approach merely as a mathematical trick, but slowly became convinced that energy in nature was indeed discrete, rather than a continuum. It also took some time before his ideas were accepted in the scientific community[66]. It was no less than Einstein who used the quanta principle to explain the photoelectronic effect, as we will see shortly.

What Einstein thought about thermodynamics

Einstein, like Planck, was very fluent in thermodynamic theory. Before 1905, Einstein published several papers on thermodynamic topics. One of these dealt with the fundamentals of thermodynamic theory [Einstein, 1903]. In this work, he studied whether the thermodynamic laws could be derived from a minimum amount of elementary assumptions. As mentioned earlier, in 1905[67] he published a study in which he explained the photoelectric phenomenon. In that explanation, he not only used the

---

[65] It's likely that Planck got his interest in black body radiation from Kirchhoff, who was his teacher. In 1889, he succeeded Kirchhoff as professor at the University of Berlin.
[66] Interestingly, Planck once remarked that a new theory gets accepted not because its opponents become convinced, but because they eventually die and new generations of scientists, unhindered by historic baggage, simply assume the theory is true (provided that it is still supported by experimental facts)!
[67] 1905 was also the year Einstein published his Special Theory of Relativity, along with his articles on the photoelectric effect, the explanation of Brownian movement, and an article where he stated his famous equation, $E=mc^2$. Because of Einstein's overwhelming amount of important material in one year, 1905 is sometimes called *Annus Mirabilis (the MiracleYear)* [Bushev 2000].

results of Planck's discrete energy packets for the black body radiation description, but fully acknowledged Boltzmann's work, calling the expression $S = k \ln W$ "the principle of Boltzmann."

How highly Einstein regarded thermodynamics can be appreciated in the following quote:

"A law is more impressive the greater the simplicity of its premises, the more different are the kinds of things it relates, and the more extended its range of applicability. (..) It is the only physical theory of universal content, which I am convinced, that within the framework of applicability of its basic concepts will never be overthrown."

Einstein is best known for his invention of relativistic theory where time is no longer invariable[68]. Less remembered is that he searched his whole life for a theory that could unify the electromagnetic theory of Faraday and Maxwell on one hand, and the mechanical theory of the material particles of Newton on the other. For instance, Newton unified the observations of falling objects on earth with the fact that the earth and planets orbited the sun. He did this by using a *single* concept – namely, gravity – to explain both phenomena. Maxwell showed that seemingly quite different magnetic and electric observations could be described by a single theory of electromagnetic waves.

As we have seen, around 1900 several outstanding physicists were working to explain Planck's black body radiation. Planck had to introduce quantum theory to explain the experimental observation of the relationship between energy and wavelength. Einstein did not like this explanation, since it introduced yet another theory rather then unifying existing theories. Einstein was convinced that the answers could be found in thermodynamics, since this theory was based on structure-independent assumptions. Indeed, the special theory of relativity can be considered as a theory of principles analogous to the theory of thermodynamics [Klein, 1967].

---

[68] Einstein worked for several years at the Swiss patent office in Bern. During that period, because of the ongoing electrification and synchronization of clocks in the cities and across the countries, many patent applications came in that proposed all sort of ingenious ways to implement the synchronization. Because of that Einstein saw of course many proposals dealing with these kinds of problems and that may have very well triggered his interest in time, see also footnote 72, [Galison, 2004].

What brought Einstein to his Special Theory of Relativity was his idea (conceived when he was 16 years old!) that the velocity of light must be the same for all observers, regardless of their respective speeds. He derived this conclusion from Maxwell's electromagnetic equations and so kept his mind puzzled for a long time. His familiarity with thermodynamic theory also gave him a lot of inspiration. We can appreciate the challenge he found from two questions (taken from the publication of Klein). In essence what the classical thermodynamic accomplishment was, was to find mathematical expressions to the dilemma:

"What must the laws of nature be like so that it is impossible to construct a perpetual motion machine from either the first or the second kind?"

This question refers to the empirical fact that perpetual machines have never been observed that could violate the first principle that energy cannot be added or destroyed in an isolated system, or contradict the second principle that entropy always increases for spontaneous processes in, again an isolated system. Similarly, while developing the Special Theory of Relativity, Einstein wondered:

"What must the laws of nature be like so that there are no privileged observers?"

This question refers to the fact that the speed of light is the same for all observers, regardless of how fast their platform (a planet, a rocket, or an angel's wings) is going. Therefore, one must derive expressions that will obey the principle of the constancy of light speed. In the same way that classical thermodynamics does not worry about *why* energy is conserved or why entropy increases, so Einstein didn't try to puzzle out *why* the speed of light was constant, but merely accepted it as fact. Once accepted, the equations that describe this assumption are pretty straightforward!

Thus, the Special Theory of Relativity can be viewed as a theory of principles analogous to thermodynamics, and not as a constructive theory – as, for instance, gravity or the kinetic gas theory[69].

---

[69] The kinetic gas theory starts with the existence of gas molecules, their continuous motion, and their finite dimensions. Then, by applying Newton's mechanical kinetic theory it is possible to derive a relation among the

This means that no model is needed (like a model of an atom in the case of quantum mechanics) in either the Special Theory of Relativity or in thermodynamics, in order to arrive at the end results of both theories. The nice thing is that both theories can live on indefinitely with little risk of needing adjustment because of new insights. That is, in fact, what we've seen: both thermodynamics and the Special Theory of Relativity have not changed since their conception. [70]

What Erwin Schrödinger thought about thermodynamics

While the quantum mechanical framework was being developed after Plank's discovery in 1901, physicists were wrestling with the dual character of light (wave or particle?). Thomas Young's double slit experiment in 1803, where interference patterns were observed, seemed to show without doubt that light was a wave phenomenon. However, Planck's interpretation of black body radiation as light quanta, followed by Einstein's explanation of the photoelectronic effect, both contradicted the light-as-wave theory. Additionally, a shocking discovery was made by Compton in 1925. Compton found that when he let X-rays (a form of light with extremely short wavelengths) collide head-on with a bundle of electrons, the X-rays were scattered as if they were particles. This phenomenon became known as the "Compton scattering experiment." At about that time, French physicist Louis de Brogly combined two simple formulas: Plank's light quanta expression ($E = h\nu$, with ν as the

---

macroscopic gas parameters: pressure, temperature, and volume. In this way a model can be built that has predictive and verifiable power.

[70] I feel that a few more words are needed here. Einstein himself pointed out in an article in 1919 in the *Times* of London that a theory of principle is based on empirical observations without the need for a particular model whereas a constructive model will first make assumptions about a fundamental structure then will built a mathematical description of that structure that hopefully will give relationships between the empirically observed parameters. In his own words: *"Thus the science of thermodynamics seeks by analytical means to deduce necessary conditions, which separate events have to satisfy, from the universally experienced fact that perpetual motion is impossible"*. Thus, classical thermodynamics can be regarded as a theory of principles, whereas statistical thermodynamics (i.e., Boltzmann approach) should be categorized as a constructive theory. In 1904 it was Poincaré who made a similar classification in scientific theories in his book *The Value of Science*.

frequency) and Einstein's famous energy-mass equation ($E = mc^2$). This led to another simple equation: $\lambda = h/mc$, with λ as wavelength. This equation really tells us that all matter has wave properties. However, since the mass, m, of most everyday visible objects is so large, their wavelengths are too small for us to notice any wave effect. But when we consider the small masses of atomic particles such as electrons and protons, their wavelengths become relevant and start to play a role in the phenomena we observe. All this brought Erwin Schrödinger to the conclusion that electrons should be considered waves, and he developed a famous wave equation that very successfully described the behavior of electrons in a hydrogen atom. Schrödinger's equation used a wave function to describe the probability of finding a rapidly moving electron at a certain time and place. In fact, the equation confirmed many ideas that Bohr used to build his empirical atom model. For instance, the equation correctly predicted that the lowest energy level of an atom could allow only two electrons, while the next level was limited to eight electrons, and so on. In the year 1933 Schrödinger was awarded the Nobel Prize for his wave equation.

Schrödinger had, as did Planck and Einstein, an extensive background in thermodynamics. From 1906 to 1910, he studied at the University of Vienna under Boltzmann's successor, Fritz Hasenöhrl. Hasenöhrl was a great admirer of Boltzmann and in 1909 he republished 139 of the latter's scientific articles in three volumes [Hasenöhrl, 1909]. It was through Hasenöhrl that Schödinger became very interested in Boltzmann's statistical mechanics. He was even led to write of Boltzmann, *"His line of thoughts may be called my first love in science. No other has ever thus enraptured me or will ever do so again* [Schrödinger 1929]*."* Later he published a book, *Statistical Thermodynamics,* and several papers on specific heats of solids and other thermodynamic issues.

The interpretation of time and its direction

What is time? [71]

A vast amount of books have been written on the interpretation of time, its connection to the fundamentals of the universe, its direction, and so on. I will not attempt here to repeat this but refer to the references in the back of the book for the interested reader and limit myself to only a few remarks.

One can say that time is truly a human invention and exists only in our minds. On the other hand, time enters in many physical formulas, the most simple one is perhaps that of speed calculated as a certain distance traveled in a certain time and is expressed in meters/second, thus being a quite tangible item.

Near the end of the medieval era we saw the invention of the mechanical clock. Before then people relied on sun clocks, which of course gave rise to differences all over the world in terms of duration of an hour, the synchronization of time, and, moreover, obviously did not work after sunset. The synchronization of time over large distances has quite some history [Galison, 2004]. There were at least two reasons why civilization wanted to have reliable time synchronization. One was the discovery of America and the other one the development of a reliable train schedule which required the synchronization of clocks in different cities. There were many occasions of colliding trains simply because of differences in time at different locations along the train rail network. Especially in France, several attempts were made to synchronize clocks; one used pneumatic systems driven by steam pressure. However, even at short distances, such as within the boundaries of Paris, this did not work at all. The use of a telegraph allowed synchronization over large distances and Britain and the USA were leading those efforts. In 1885 there was an international conference that agreed on a standard time (Greenwich Time) and the division of the world into 24 time zones. Humans, it seemed, captured time and put it in a convenient box.

---

[71] Saint Augustine wrote, "I know what time is, if no one asks me, but if I try to explain it to one who asks me, I no longer know." (from *Confessiones*).

Einstein's interpretation of time

However, Einstein's relativity theories would blow the lid off traditional notions of time, because they led to such bizarre conclusions[72]. For instance, imagine two observers, A and B, both equipped with very accurate clocks showing the exact same time. A takes off in a starship and travels at a velocity close to the speed of light. During their separation, both observers watch their clocks and do not see anything abnormal. But when A returns and rejoins B to compare clocks, both notice that A's clock is running behind. (This assumption actually was proven in a recent earthly experiment, as we'll see soon.)

The actual results in mathematical form of Einstein's relativistic theory are not too difficult to understand. Let's have a look at a famous relativistic formula, the one for the dilatation of time and contraction of distance. For time, this is:

$$Time = \frac{Time\_at\_rest}{\sqrt{(1 - v^2/c^2)}} \quad (4.1)$$

where "*Time*" is the time in the system that travels with a velocity v relative to the system at rest. This famous expression describes the important relativistic conclusion that time is not constant, but increases when speed approaches light velocity, c. From this formula comes the

---

[72] One of the earliest involvements of Einstein with the concept of time was the rather fundamental problem of how to synchronize clocks in different locations [Galison, 2004]. This was among others triggered by the fact that Einstein was studying how physical laws would be perceived by observers in different locations at different speeds relative to each other. This can be illustrated by a very simple example. Imagine we have an observer A in point A sending a light pulse at time t=0. An observer B in the same location as observer A will agree with observer A that the light pulse was sent at the same time. An observer C at a certain distance d from observers A and B will see that light pulse only at a time t = d/speed of the light. Thus, observer C will not see the light pulse at the same time as observers A and B and therefore not all events will occur to all the observers at the same time. The same problem plays in the synchronization of clocks. Einstein solved this by correcting the time of the remote clocks for the time it takes the synchronization signal to go from the reference clock to the remote clocks.

famous conclusion that time is not the same everywhere in the universe; it slows down noticeably when you start to travel at velocities near the speed of light. Now, let's look at another relativistic formula for length contraction at high speed v:

$$Length = (Length\_at\_rest)\sqrt{(1 - v^2/c^2)} \qquad (4.2)$$

which closely resembles the formula for time. Equation (4.2) shows that length will contract at high velocities.

In Table 4.1, we can see how time becomes longer and length becomes shorter when velocity increases.

Table 4.1 Relativistic Time and Length		
Speed in km/hour	Relativistic Time	Relativistic Length
0	1.00	1.0000
100	1.00	1.0000
1000	1.00	1.0000
10000	1.00	0.9994
100000	1.06	0.9428
200000	1.34	0.7454
250000	1.81	0.5528
275000	2.50	0.3997
290000	3.91	0.2560
295000	5.50	0.1818
299000	12.26	0.0816
299900	38.73	0.0258
299990	122.48	0.0082
299999	387.30	0.0026

The interesting thing about Table 4.1 is that only at very high speeds do we see a noticeable relativistic effect. Up to about 10,000 km/hour, no effect is visible. And so we can understand how, in everyday situations, we can still measure the world with Newton's formulas.

Thus from the relativistic formulas, we learn that time is not a constant given but differs between observers, and is affected by the systems in which they live. This is not just hypothesis – the concept was verified by the famous Hafele and Keating experiment in October 1971.

In that test, four cesium atomic clocks were flown on commercial jets eastward and westward around the world. The traveling clocks then were compared to reference clocks at the US Naval Observatory. The jet-set clocks were found to have lost 59 nanoseconds on their eastward trip, and gained 273 nanoseconds on their westward excursion very much in agreement with the values predicted by Einstein's theory! (The difference between the westward and eastward time lags is caused by the rotation of the earth).

The influence of modern physics on thermodynamics: does relativity change entropy?

We can't help but wonder how quantum mechanics and relativity will ultimately impact the laws of thermodynamics. So far, quantum mechanics seems to have had little impact, other than the effects we saw in Chapter 3 where we discussed MEMS and other small devices. Since thermodynamics looks only to macroscopic parameters such as pressure and temperature, atomistic dimensions appear to play no role. On the contrary, thermodynamics works only for large numbers of atoms or molecules. It was exactly for that reason that Boltzmann introduced the idea of statistical thermodynamics.

However, things are a bit more complicated with relativity. For the First Law of Thermodynamics, relativity is irrelevant: conservation of energy holds in both domains. For relativity's effect on the Second Law, however, we need to go into some detail. Several scientists, including Planck, von Laue, Einstein, and Tolman developed relativistic formulas for both heat and temperature [Schlegel, 1975]. Their results can be summarized as follows.

For heat:

$$dQ = dQ_0 \sqrt{(1 - v^2/c^2)} \qquad (4.3)$$

and for temperature:

$$T = T_0 \sqrt{(1 - v^2/c^2)} \qquad (4.4)$$

Here, Q_0 and T_0 are heat and temperature in a physical system at a speed v relative to the observer at rest, for whom heat and temperature become Q and T, respectively.[73] Now, we look again at the familiar expression, $dS = dQ/T$. When we divide equation (4.3) by equation (4.4), we find that all the relativistic terms cancel:

$$dQ/T = d\, Q_0/T_0 \qquad (4.5)$$

and therefore $S = S_0$. Thus it appears that the entropy for both the observer at rest and the observer at velocity v is the same. This is an important conclusion, since it really means that there is no need for something like a relativistic thermodynamic adjustment. This result is easy to understand using the Boltzmann interpretation of entropy. Entropy is a measure of the probability that a certain macrostate can be realized (the more microstates are possible in a given macrostate, the more probable that macrostate becomes and the higher the entropy). Indeed, the amount of possible microstates apparently will not change as the speed of its system changes. Therefore, we would not expect entropy to be affected by velocity.

---

[73]However, many more discussions by other scientists have taken place since Einstein and Laue around 1950 proposed these formulas and it is fair to say that new insights are still developing and deviate from the above but fall outside the scope of this book.

PART II

Entropy and Our Society, Our Culture, Our Planet, and Our Universe

"The entropy principle defines order simply as an improbable arrangement of elements, regardless of whether the macro-shape of this arrangement is beautifully structured or most arbitrarily deformed."[74]

---

[74] Rudolf Arnheim in "Entropy and Art, an essay on disorder and order", University of California Press, Berkeley and Los Angeles (1971)

5
Entropy, the Economic Process, and the World's Environmental Problems

Humanity has received many warning signs about economic growth and the associated negative impact on the environment. Among the first monitors was Thomas Robert Malthus, who in 1798 published "An Essay on the Principles of Population." In that essay, he predicted that world population growth would outpace economic growth, which would be constrained by the limited amount of arable land. Consequently, he predicted this imbalance would lead to social misery. Nearly two centuries later, economist Nicolas Georgescu-Roegen in 1971 pointed out that economic processes, as generally assumed, are not cyclical at all, and will in the long run lead to exhaustion of the world's natural resources.[75] Then in 1972, the Club of Rome published the shocking report, *The Limits to Growth*, followed by another well known book in 1989 by Jeremy Rifkin, *Entropy Into the Greenhouse World*. In this chapter, we will show how the Second Law impacts the economic processes that affect our environment. Although in the last few years,

---

[75] The Dutch-born economist Tjalling Koopmans (1910-1986) wrote from 1951 onward about limited natural resources and the irreversible nature of the economic process. However, it was Georgescu-Roegen who introduced the entropy concept to economic theory. Likewise, Robert Ayres in 1969 introduced the Law of Conservation of Matter to economic modeling.

ecological economists have paid increasing attention to the true impact of entropy on economics, the concept remains controversial within the mainstream economic community.

General environmental trends

Over the last 50 years or so, awareness of the impact of human actions on the environment has increased. For at least 200 years, humanity has created large-scale economic activities in order to meet market demands. The overarching Western economic model is that of continuous growth, predicated on the idea that growth will come automatically from technological "progress." There are many reasons or justifications for this glorification of growth, which perhaps is an inherent property of the economic process. For instance, some believe that growth is needed to eradicate poverty and maintain full employment. Some governments use economic growth to prevent their repressive political systems from being undermined. Others believe growth is the only way to keep long-term profits up, since in a steady state they tend to decline. Whatever the merits of these beliefs, however, it's a fact that human economies impact the environment, and entropy describes that impact. But first, let's consider a few facts about our economic system.

Undoubtedly, one of the most important factors that determines world economic activity is population. Figure 5.1 shows the growth of world population over time. Global population increased dramatically from 2.3 billion in 1950 to 5.7 billion in 1995. From there, depending on which growth scenario (low, medium, or high) is accepted, world population goes to as high as 11.2 billion people in 2050. With the proliferation of economic growth in developing parts of the world, one would expect that global economic activity also will increase greatly, as illustrated in Figure 5.2. The world's economic output increased about threefold from 1970 through 2002, while world population grew approximately 50% in that same period. Such increased economic activity obviously needs natural resources such as energy (mainly from fossil fuels) and materials (for example, wood, iron, and copper). It should be noted here that natural resources can be characterized as renewable and non-renewable. Renewable resources, such as solar, wind, and hydro energy, are renewable in the sense that future generations and we theoretically have continuous access to them, although output at any one time is limited. Non-renewable resources include all fossil fuels and

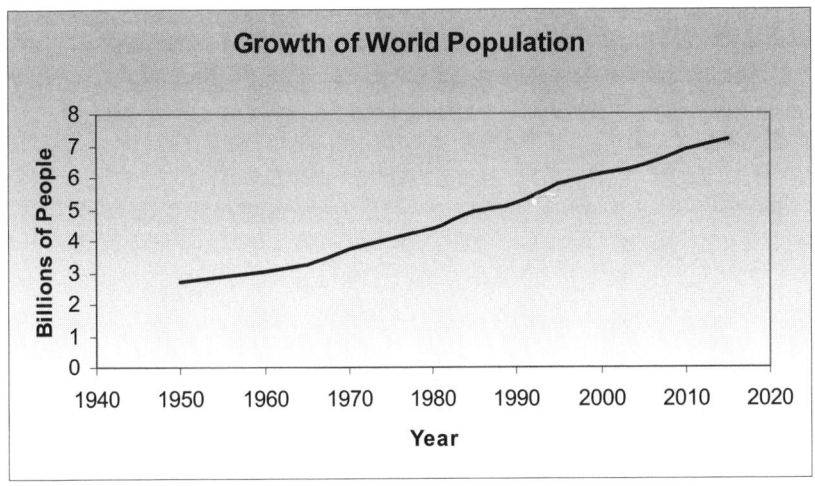

Figure 5.1 Development of World Population over time (Source: US Bureau of the Census).

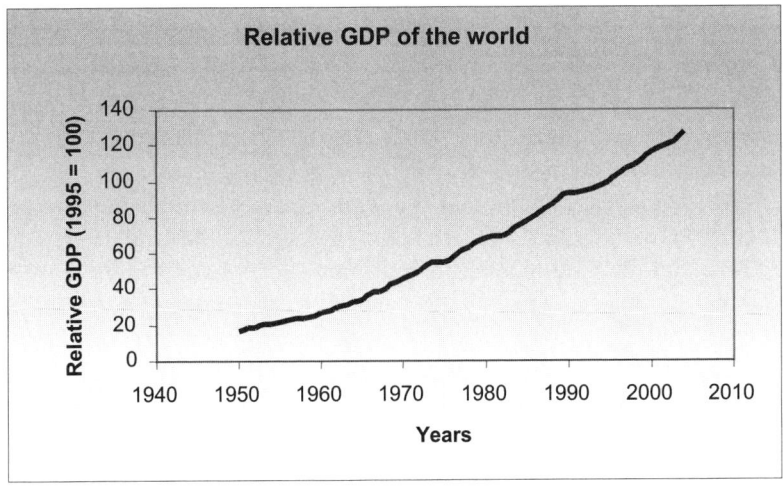

Figure 5.2 World Gross Domestic Product (Source: World Trade Organization).

natural materials, such as minerals and metals. These natural resources are irreplaceable, and what we consume today will simply not be available for future generations.[76]

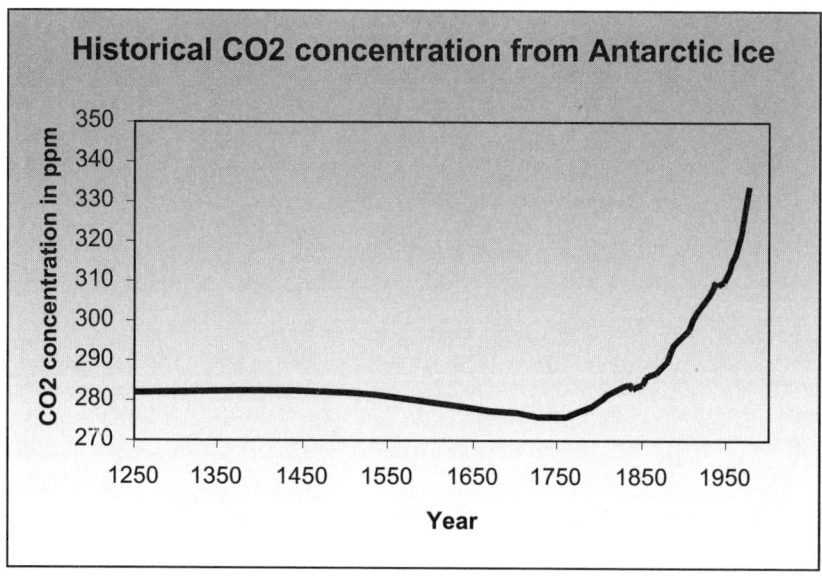

Figure 5.3 Concentration of CO_2 from 1250 onwards (data source Etheridge et. al., Division of Atmospheric Research, CSIRO, Aspendale, Victoria, Australia).

Apart from the need for natural resources (discussed later in this chapter), economic processes also bring waste production. One of the most intensely discussed topics in this area is air pollution, notably the impact of greenhouse gases on global warming, the destruction of the ozone layer, and the phenomenon of acid rain. Greenhouse gases prevent sunlight reflected from the earth's surface from escaping into space. These gases include carbon dioxide (CO_2), chlorofluorocarbons (CFCs, $CFCl_3$), methane (CH_4), and nitrous oxide (NO_2). From concentration

---

[76] Some would say recycling of materials would put these in the category of renewable resources. We'll see later that this approach is very limited. Apart from the fact that recycling can never reach 100%, the truth remains that associated entropy production is irreversible.

5: Entropy, Economics, and Environmental Problems 99

trend graphs, it is undeniable that air pollution is present today – and in fact has been with us for centuries! In Figure 5.3, we can see how the atmospheric concentration of CO_2, produced from the combustion of fossil fuels, has increased over the years [Etheridge, 1996 and 1998]. Before 1750 the CO_2 concentration hovers around 280 ppm but after 1800, we see a sharp increase, which of course coincides with the beginning of the Industrial Revolution and its associated use of wood and coal as fuel.

The increase of CO_2 (and of CFCs since about 1960) has been linked to global warming (also called the greenhouse effect), although scientists still do not agree about the exact mechanism of this phenomenon. Nevertheless, global warming is expected to raise average temperatures between 2 and 4^0 C over the next 100 years, with ominous consequences – rising ocean levels, for example. Meanwhile, CFCs are degrading the ozone layer in the stratosphere at heights between 12 and 70 km.

The ozone layer, an ancient protector of life on earth, is created when oxygen in the upper atmosphere (about 150 km high) absorbs energy from sunlight, which then breaks the oxygen molecules (O2) apart into single, reactive atoms. These atoms then recombine with oxygen molecules to form ozone (O3). Ozone absorbs ultraviolet (UV) radiation from the sun and so protects plants and animals from an overdose of UV. CFCs are used in refrigerator systems in cars, homes, and industry. These chemicals eventually break apart, releasing chlorine that reacts with ozone to break it up. The result is that since about 1985, an increasing "hole" has developed in the ozone layer. Although measures have been taken to ban CFCs and replace them with less harmful agents, it must be noted that CFCs can persist in the atmosphere for up to 400 years. Therefore, even a strict ban on CFCs cannot immediately reverse the damage that is now occurring to the ozone layer

A group that has definitely put the environmental issue on the political agenda is the Club of Rome. Aurelio Peccei, an Italian industrialist, and Alexander King, a Scottish scientist, felt that something needed to be done to protect the earth's environment from industrial destruction. In 1968, they called for an international meeting of 36 European scientists in Rome. In 1972, this group published a report, *The Limits to Growth,* which described the relationship between economic growth and damage to the environment. This report immediately had a tremendous impact on the public and political discussion of environmental issues (or perhaps it finally started a much-needed broad, environmental discussion in society)[77]. The group developed a dynamic global model in which five major trends were incorporated: increase in industrial activity, escalating growth of world population, depletion of natural resources, extensive malnutrition, and finally, a fading environment. The model's shocking conclusion was that, if nothing were done, exhausted natural resources would cause a global collapse well before the year 2100. Also, different economic scenarios and their impacts on resources were included in the report. The grim outlook was that even if we could add resources or implement new technologies, the collapse could not be avoided. This happens because the fundamental problem is an *exponential growth in a finite and complex system*. The group proposed a way out of this situation by going to a stable state where industrial output per capita is frozen at the 1975 level, and global population growth ceases. Today, the Club of Rome continues to exist, has members from 52 different nations, and still issues reports on environmental matters.

A slightly different way to put the ecological challenge in perspective is by introducing the concept of *carrying capacity* [Rees, 1996]. Carrying capacity is the maximum persistent load that a given environment can support for a certain purpose (for instance, to continue human civilization). Human civilization needs energy, minerals, wood, food, etc. The carrying capacity can be calculated in terms of how much land area is needed to support our civilization. The resulting calculations of Reese and his coworkers yielded remarkable results. For instance, in 1995 there were 1.5 hectares of arable land per capita available if one

---

[77] Apart from the Club of Rome, additional events contributed to the environmental awareness of a wider audience. To name a few: The Vietnam War from 1950 till 1975, where Agent Orange had a devastating impact on the tropical forests; the foundation of Greenpeace in 1971; the release of dioxin in the Italian town of Sowezo in 1977; the nuclear disasters in 1979 at Three Mile Island and in 1986 at Chernobyl; the methyl isocyanide leakage in Bhopal in 1984, and the oil spillage from the Exxon Valdez in 1989 all had deep impact on public opinion. The general perception was that environmental care was wrongly entrusted to governments and corporations. Nevertheless, some effective actions were taken, such as the establishment of the Environmental Protection Agency in 1970 and the US Clean Air Act in 1990. Other results included the ban on DDT in agriculture in 1972, reduction of carbon monoxide emissions from cars by use of catalysts in the exhaust systems, and reduction of overall lead emissions by a factor of 40 between 1970 and 1990.

5: Entropy, Economics, and Environmental Problems   101

takes into account the total available productive land on earth. However, the citizens of Vancouver, BC in 1991 needed about 4.2 hectares per capita to keep up with their consumption – which means that the city's 472,000 inhabitants required a total of 1.99 million hectares of land, compared to the urban area of only 11,400 hectares. An even more dramatic example is found in the Netherlands. Because of its large population density, this country needs to draw resources from an area of land that is 15 times (!) bigger than the entire country. Looking at the situation globally, for the rest of the world's population to live at the same level as North Americans do today, humanity would need another habitable planet at least as big as earth in order to support its overall lifestyle.

The considerations above are only meant to set the scene for the discussion in the rest of this chapter. Many more details about ecology can be found in a vast array of literature, but that exploration is beyond the scope of this book. However, the picture is clear: economic activity does impact our environment. In the following pages, we will discover how the concept of entropy can help us in understanding the interaction between the economic process and the depletion of terrestrial resources, as well as mounting pollution.

How entropy plays a role in the economic process and re-defines concepts such as efficiency and sustainability

Relationship between thermodynamics and economic processes

What is the "economic process?" Many definitions can be found, but here we would like to use the description by Nicolas Georgescu-Roegen[78] in his seminal book, *Entropy Law and the Economic Process*: The main objective of the economic process is to ensure the long-term existence of the human species. Another way of putting it is to say that

---

[78] Nicolas Georgescu-Roegen was born in Romania in 1906. Although he obtained a PhD in mathematical statistics, he familiarized himself with the new field of economics during a stay at Harvard University. In 1948, he fled Romania because of the Communist regime and returned to the US to take a position at Vanderbilt University, where he published many studies in economic science.

the output of the economic process is designed to maintain and increase the enjoyment of life. He defines three important partial economic processes: agriculture, mining, and industrial manufacturing.

In Chapter 2, we mentioned that Sadi Carnot, the young French military engineer, was one of the first scientists to apply science to industrial problems. He wanted to understand the efficiency of the steam engine (i.e., how much coal was needed to deliver an amount of work) and accomplished that, as we saw in Chapter 2, by a painstakingly detailed analysis of the cyclic nature of the steam engine. Only after that analysis did it become clear that the efficiency was only dependent on the temperature *difference* between the steam in the boiler and the vapor in the condenser. This ended an earlier belief that the efficiency of the steam engine was not limited at all. We learned that of the available heat at high temperature, only a portion could be used to perform work, and that the rest of the heat was wasted at the lower temperature of the condenser, and so could not be used anymore to deliver work.

Thus, in applying heat to do work, we don't actually consume energy (or heat, for that matter). Indeed, the total amount of energy stays exactly the same. It is the qualitative aspect of the energy that changes: it transforms from the available form (i.e., available to perform work) to the unavailable form (cannot be used anymore to perform work). Now we come to the heart of the matter. In typical economic models, the economic process is pictured as cyclic. This image is triggered (among other reasons) because money, a not unimportant item in an economy, goes basically from one hand to another without ever disappearing. Without further knowledge, one sees some parallels between money and energy[79]. As mentioned, energy stays constant throughout the entire economic process, so why would we be worried? But we know better; we know what the fallacy is. We also know that the entropy of an isolated system increases and that this increase is irreversible or even irrevocable. Thus, all this boils down to the Entropy Law, that says that there is only one direction we can go on this planet and that there is no way back. Therefore, the assumption that the nature of the economic process is cyclic is a serious misjudgment. More about this in the sections that follow.

---

[79] One can argue that just as thermodynamics has its law of energy conservation, economics has a law of the conservation of money [Mirowski, 1988].

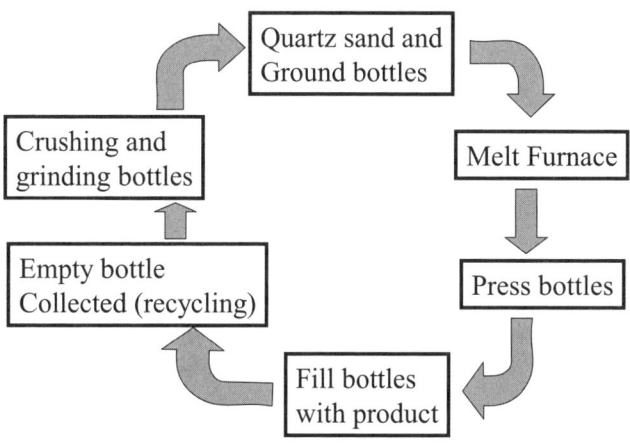

Figure 5.4 Life cycle of a glass bottle.

Example of an economic process and the Entropy Law

In this section we want to analyze in more detail how the connection between the economic process and the Entropy Law comes about. To do this analysis, we need of course a description of the economic process. Although (as mentioned above) all economic processes can be categorized as agriculture, mining, and manufacturing, we cannot possibly analyze them all because of the immense amount of detail involved. For that reason, we'll concentrate on a single manufacturing example: the life cycle of glass. Why glass? Well, I've always been impressed and intrigued by the art of glassblowing. But there are more practical reasons, too. Glass plays a role in many aspects of everyday life – tableware, windows, glasses, television tubes, and so on. And although glass itself does not contaminate our environment, the glass-making process certainly does, as we will see. Finally, glass is one of the best examples of a material that can be recycled almost forever, but we will see that its recycling behavior is somewhat misleading as well. Most glass production occurs in two forms, glass containers and flat glass. For this analysis, we will consider only glass bottles.

The life of a glass bottle starts with the gathering of the raw material, called silica (a chemical compound of silicon and oxygen, SiO_2) that can be found as quartz sand, or in the case of recycling, crushed glass bottles. Quartz sand is in fact already glass, but in the form of very fine grains. The quartz sand is fed into a furnace for melting; sometimes small amounts of other materials are added to change the properties of the glass (for instance, lead oxide to produce lead crystal glass, iron to make brown or green colors, and cobalt salts to produce green or blue tints).

Now let's go a step further and look at how the steps described in Figure 5.4 impact entropy in a qualitative sense (that is, the entropy increases or decreases). We'll do this at the completion of each step, except for the raw materials step, since that is our starting point. Also, we will break down the entropy changes throughout the cycle for the glass components and for the impacted environment at all stages. We'll call that environment the "World" and we'll define the World's boundaries so that outside them, no further impact of our manufacturing process can be detected. In other words, we'll have an isolated system where we can see not only the entropy changes of the glass-making process, but also the impact of that process on its environment. We will arbitrarily assign an entropy to the raw materials and the World and estimate increases or decreases from that point onward.

The next step is the melting furnace, which can run as hot as 1700^0C. The silicate molecules are now in liquid state, and we know from the Boltzmann formula that the entropy will increase, because we now have more microstates available for the silicate molecules. But what will happen with the entropy change in the World? Well, we will need lots of energy to heat the furnace and keep it at high temperature. For that, we can burn fossil fuels, which will cause another entropy increase (as we have seen in the steam engine discussion) by producing large quantities of CO_2 and consuming big chunks of energy. In fact, each ton of glass melted in the furnace will use about 4 giga-joules! Thus, we expect that the entropy will increase for the glass as well as for the World. In Figure 5.5, we can see this as an increase in the entropy bars.

Once the glass is melted in the furnace, the next step is to pour gobs of the liquid glass into a mold, where the glass will be shaped by mechanical action or compressed air into its final form as bottles. A well controlled cool down step follows to ensure that all the stress in the bottle is relieved and no cracks occur. As the glass goes from liquid state to bottle form, the entropy will drop because the material is going to a less chaotic state. But with respect to the World, we need energy to compress the glass into molds, which will increase the entropy. In addition, cooling down the glass will release heat to the World, leading again to an entropy increase. We estimate that the entropy decrease of the bottles is less then the entropy increase in the World, and so the net result will be, once again, an entropy increase of the entire isolated system.

The next step is to fill the bottles with products such as water, beer, sodas, medicine, and so on. (We do not expect any substantial entropy change from how these products will be used.) Let's assume next that all the empty bottles will be recycled[80]. Making a bottle out of recycled glass, which does not reduce the quality of the glass, saves about 30% on the energy bill – or enough per bottle to let a 100-watt bulb shine for an hour! Recycling also brings a 20% reduction in air pollution. How about the entropy balance? Well, to do the recycling, people will have to drive to recycling centers to drop off their empty bottles, and trucks will have to transport the bottles to the glass factory. The entropy of the glass will not change that much, but spent fuel from the car and truck trips will increase the entropy of the World and therefore the entropy of the entire isolated system.

In the glass factory, the bottles now must be crushed. We could argue that the entropy of the glass will increase a bit because the state of the crushed glass is more chaotic than that of the bottles, but the difference is negligible. However, the entropy of the World will increase again because we will use electric power to run the crushing machine, This in turn will burn more fossil fuel and lead to an entropy increase, as indicated by the higher bar in Figure 5.5.

At this point we are back to the raw material for bottle production: the crushed glass will be mixed with silicate sand to feed the melt furnace, and the whole process will start again. Thinking only of the glass-making process, it looks like we have a closed cycle, as pictured in Figure 5.4. But when we examine Figure 5.5 from an entropy point of view, there is nothing like a closed cycle: the entropy has increased since we started production. We have to conclude that recycling glass bottles is not at all a cyclic process!

---

[80] In fact, in countries such as Finland and Switzerland, the recycling rate for bottles is about 90%! However, not all countries are that far along. In the UK, for instance, only 35% of glass bottles are recycled.

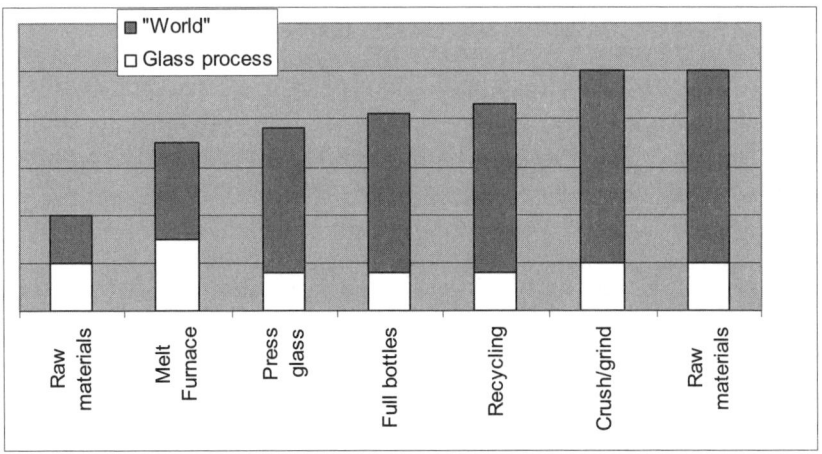

Figure 5.5 Entropy changes in the glass life cycle.

A plea for a redefinition of efficiency and sustainability

The main point Nicolas Georgescu-Roegen made was that economists did not pay sufficient attention to the side effects of economic processes. Often, the situation was pictured such that there were no issues in obtaining world resources and that "better" technology eventually would solve all environmental problems. However, our analysis of the glass making process powerfully shows how wrong this worldview is! Yes, from a purely materials and theoretical point of view we can continue with glass recycling indefinitely (at least in theory, more about this later), but we know now that from an entropy perspective, the process constantly increases the entropy of our world. Of course, the world's entropy will increase as part of nature; but the point is that unchecked human industry will accelerate it at a much greater rate.

We have now established that the economic process transforms available energy and the world's resources into a situation of high entropy, and at a much faster rate than natural processes would drive. (The increase in entropy is partly counteracted by a decrease of entropy caused by the solar radiation; see also Chapter 6). This high entropy situation can be described as a continuous formation of waste by the economic process. This quite fundamental and mostly hidden mechanism is not recognized by society or mainstream economists. This deficiency

5: Entropy, Economics, and Environmental Problems

has everything to do with the traditional definition of efficiency. Efficiency typically is seen as how much output a certain process can generate versus the amount of input required in terms of headcount, raw materials, or capital (among other resources). All effort is then focused on increasing output and decreasing input. This is called efficiency improvement but sometimes is also termed economic development. However, if we want to account for world resources, we must simultaneously consider *how to decrease the amount of entropy production*. Thus in our bottle-making process, that means considering not so much the entropy changes in the glass process itself, but also the entropy increase in the World. Given this approach, it would make sense to define something like an *entropy efficiency*, or the total amount of entropy produced per bottle. If we would incorporate that kind of efficiency into our general approach to improving processes, we would truly achieve a sustainable economy.

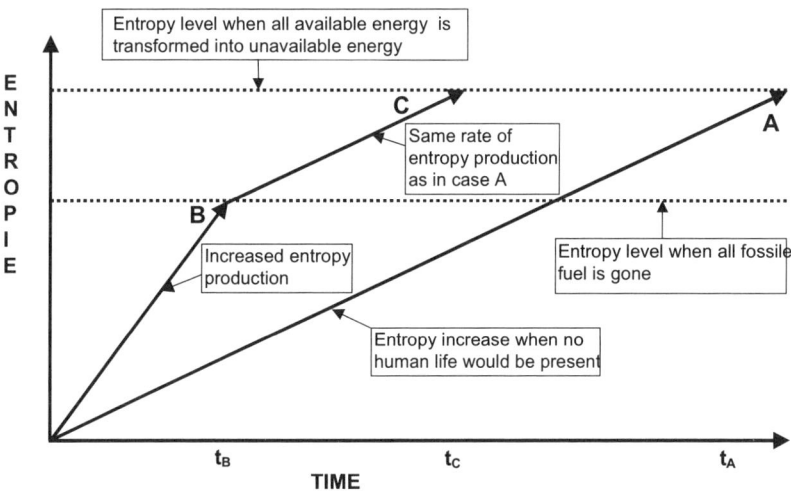

Figure 5.6 Simplified schematic of entropy production in the world.

Let's elaborate a little bit more on the issue of entropy production. We have seen in Chapters 2 and 3 that the entropy of an isolated system increases. We have also seen that ultimately the entropy will reach a maximum value and that everything will come to standstill. Of course, we hope that it will take quite some time before we reach that situation. The moment our economy comes to a standstill will be reached long before the entropy of the world assumes that maximum value. We also know that we need to transform energy every day to keep our economy going. But when all available energy has been transformed, there will be no human life (at least not as we know it today) left on earth and the rate of entropy production will slow down. In Figure 5.6, we have tried to capture the situation in a simplified version but it will do to make our point. What we see is how the entropy in the world develops with time.

Line A is our reference case. That is the rate of entropy production in the world if no human life (or any other "civilized" life form) is present. The entropy will steadily increase per the Second Law of Thermodynamics[81], and will go on till we reach time t_A, where finally all available energy has been transformed into unavailable energy. We assume that it will take hundreds of millions of years to get there.

In case B, represented by arrow B, there is human life on earth, including the entropy-inefficient economic processes described above. The rate of entropy production is now increased compared to case A, and will go on until we have transformed all fossilized energy and other natural resources on the planet. At that time, represented by t_B, virtually no economic (industrial) activity is possible, with serious consequences for human life. It is clear that t_B can be much shorter than t_A. This shows that we leave much less time for future generations to be able to live on earth regardless of all the recycling we do! We know now that the entropy cannot be recycled. After t_B, the entropy production will continue at a rate approximately the same as for case A, since no significant human activity will be present. The only difference is that time t_C, where the maximum entropy is achieved, occurs in a much shorter time than in case A.

Despite the fact that the concept of entropy reveals some severe limitations of our current way of life, it is fair to say that the concept of entropy has never caught on in mainstream economics literature. One can find both publications in favor or against the use of entropy in economic analysis (see, for instance, Krishnan, et al. 1995). The ideas of

---

[81] In fact, our planet is *not* an isolated system but rather a closed system. For instance a continuous stream of energy is transferred from the sun to the earth. This influx of solar energy actually causes the entropy increase by the spontaneous natural processes to slow down. See Chapter 6 for more details on the entropy development of the earth.

Georgescu-Roegen, however, did find a lot of supporters in the ecological-economic literature. Mainstream economists believe that there will be a technological solution for the energy and materials scarcity problem, despite the entropy arguments of the former group. And it is truly difficult to predict the future of technology – as when Western Union declared in 1876, *"The telephone has too many shortcomings to be seriously considered as a means of communication. The device has no value."* Indeed, it has been widely proposed to convert the sun's energy into electrical power by utilizing solar devices based on semiconductors. This is an option to which we should give more thought, since the sun is a nearly inexhaustible source of energy. The energy that is radiated by the sun onto the surface of the US is 50,000 Quads per year[82]. Per year, the US transforms about 100 Quads of available energy into non-available energy. Clearly, on paper we have a tremendous potential, a real bonanza. However, we must make a few caveats here. First, the conversion efficiency of a solar cell is at best 15% (although technology progress could raise that number by a factor of two). So the potential 50,000 Quads are reduced to about 7000 available Quads. Second, we cannot (and don't want to) use the entire surface of the US for solar cell panels, so let's say we use only about 1% of the available land area. Now the 7000 available Quads diminish to about 70 accessible Quads. Even so, that figures is still comparable to the yearly US consumption of energy! Thus, solar energy could become a viable renewable power source that would make our society more sustainable. See Appendix VIII for a further discussion on this topic.

Transformation of terrestrial resources from available to non-available

An adult human must transform an average of about 8000 kJ per day to survive physically. However, if we include all the other energy sources used to produce cars, electricity, processed foods, and other products for everyday living, each of us transforms about 800,000 kJ a day![83] Since everything in nature, even matter, is in essence energy (through the laws

---

[82] 50,000 Quad is about $1.4 \cdot 10^{16}$ kWh. A "quad" is one quadrillion ($1*10^{15}$) British Thermal Units (BTUs). One million ($1.0*10^6$) BTUs = 90 pounds of coal or eight gallons of gasoline (from: http://rredc.nrel.gov/tidbits.html#quad)

[83] See Rifkin in *Entropy, into the greenhouse*, 1981

of relativity, discussed in Appendix VIII) we should not be surprised that the Second Law of Thermodynamics also rules economic processes. There is no escape: every time we transform energy, we turn a certain amount of it into unavailable energy. The Entropy Law describes the amount that becomes unavailable.

Thus while the total amount of energy will remain constant, does the same hold true for materials? For instance, let's consider iron. Iron is a stable element, which means that the total number of iron atoms on earth will not change[84]. Iron is mined from areas rich in iron ores, and then through a proper process is liberated from its bonds with other elements such as oxygen. Then the iron is used in all kind of goods, from screws to automobile bodies to train tracks. After these products have served their purposes, it would seem possible to recycle 100% of the iron — but in fact this is not true at all. For instance, in the case of train tracks, every time a train runs over the tracks or brakes on them, it wears down the rails and throws iron atoms into the air. This is a process comparable to gas diffusion, where gas atoms or molecules always will try to occupy the maximum space available. (This is another instance of entropy increase, as we saw in Chapter 2.[85]) Thus, a certain amount of productized iron cannot easily be recycled, and in fact is lost to the economic process. This loss is very comparable to the transformation of available energy to non-available energy[86].

Summary of entropy and economy

In the economics literature, one can find two opposing points of view: mainstream economists who believe that technological innovation will solve the degradation in quality of both energy and materials and that therefore growth can go on forever; and biophysical economists, who use the thermodynamic laws to argue that mainstream economists do not incorporate long-term sustainability in their models. For instance, the costs to repair the ozone hole or to mitigate increasing pollution are not

---

[84] Unlike radioactive materials such as radium and uranium, which gradually degrade into simpler elements.

[85] A further discussion of how entropy increases because of diffusion is available in Appendix V.

[86] An attempt to assign a quantitative entropy amount to the liberation of a metal from an ore has been done by Faber, et al., 1983.

accounted for in mainstream economic assessments. We saw how industrial and agricultural processes accelerate the entropy production in our world. Entropy production can only go on until we reach the point where all available energy is transformed into non-available energy. The faster we go toward this end, the less freedom we leave for future generations. If entropy production were included in all economic models, the efficiency of standard industrial processes would show quite different results. Even if there were no humans on this planet, there would be continuous entropy production. So from that point of view the ecological system is not perfect, either; even the sun has a limited lifespan. The real problem for us is that, in our relentless effort to speed things up, we increase the entropy production process tremendously. In fact, you can see some similarity between economic systems and organisms: both take in low entropy resources and produce high entropy waste. This leaves fewer resources for future generations.

Although recycling will help a lot to slow down the depletion of the earth's stocks of materials, it will only partly diminish the entropy production process. So whenever we design or develop economic or industrial processes, we should also have a look at the associated rate of entropy production compared to the natural "background" entropy production. We have seen that for reversible processes, the increase in entropy is always less than for irreversible processes. The practical translation of this is that high-speed processes always accelerate the rate of entropy production in the world. Going shopping on your bike is clearly a much better entropy choice than using your car. You could say that the entropy clock is ticking, and can only go forward!

6
Energy, Entropy, Life, and Heat Death

The contradiction between the thermodynamic push for chaos and the tremendous degree of molecular and biological organization

As a boy, I earned money during summer vacation by working for farmers near the village where I lived. With that money, I bought a microscope and started studying all kinds of specimens, such as bacteria in water and cells from onion skin. I was immediately impressed how structured the organisms were, and spent hours observing the world unraveled by my microscope. And indeed, is it not almost unbelievable how a newborn child will almost automatically develop from a fertilized egg cell? That brings me to the question: how is it possible that despite the push for chaos by the Second Law nature manages to create extremely well organized forms of life? For instance, the existence of DNA, with its almost unimaginable degree of coding complexity, is mind-boggling. To answer this apparent paradox, we first have to define properly the boundaries of the system we are considering. When we do this, we come to the conclusion that the creation of life occurs at the expense of available energy elsewhere in the isolated system.[87] In other

---

[87] To recap: an isolated system is a self-supporting system that does not exchange energy or material with its environment. In studying the entropy of a

words, creation of life needs lots of energy from other parts of the system, and the overall system entropy will still increase, even as the entropy of the organism decreases (as long as it is alive).

An interesting view on the relationship between thermodynamics and life was given in 1944 by the great physicist Erwin Schrödinger in his book, *What is Life?* Being a well established physicist (see Chapter 4), Schrödinger moved into the boundary between physics and biology; he believed that the new quantum mechanical physics and the concept of entropy based on statistical mechanical concepts could help enlighten certain issues in genetics. He pointed out that in order for thermodynamics to work, the system under study must not be too small, or statistical variations will become noticeable. What is too small? As soon as the system dimensions become of the order of, say, several million atoms, we are in the "danger zone." One of the best examples of this "danger zone' is the Brownian movement that we already discussed in Chapter 3. There we saw that the random collisions of water molecules with pollens do indeed result from time to time in a net displacement of the pollen. The point Schrödinger was making is that at cell dimensions, and certainly at chromosome scales, we are clearly in the danger zone, and violations of the Laws of Thermodynamics perhaps can be observed. In the sections that follow, we will examine these situations closer.

Chaos and life

Let's begin with two main conclusions from Chapters 2 and 3. In Chapter 2, we saw that isolated systems, when left to themselves, will always increase their entropy to a point that the entropy reaches its maximum value. At that value, the system will no longer change its properties, and everything (at the macroscopic level at least) comes to a standstill. One also can say that the system has achieved equilibrium. A simple system

---

mouse, for example, we would define the system not only as the animal, but also the immediate environment with which it exchanges energy such that the system including the mouse would become completely self-supporting, without any need to exchange energy or materials with the environment outside this system. This implies typically that we must position the boundaries of our system large enough. In the case of the mouse, defining the system as the entire planet is probably overkill whereas defining the system as the mouse cage would be too small.

6: Energy, Entropy, Life, and Heat Death   115

would be an ice cube floating in a cup of water. If we do nothing, the cube will melt. After it is completely melted and the cube has turned into liquid water, no further changes are expected.

In Chapter 3, we discovered that entropy can be interpreted in terms of the amount of disorder of a system: $S = k ln W$, where W is a measure of the disorder[88]. The more chaos (or disorder) in a system, (or more precisely, the higher the number of possible microstates), the larger W will be and the larger the entropy will become. We learned that the fundamental reason for the entropy increase of isolated systems is actually a continuous drive of that system toward more disorder. Or to say it more elegantly, the system strives toward the most likely macrostate. The tendency toward disorder is fueled by the heat motion of atoms and molecules.

Thus we find that (lifeless) nature tends to chaos. However, what can we say about life? The smallest unit of life is a cell (or bacterium, for that matter)[89]. A cell is the smallest self-sustaining organism. Larger assemblies of cells become plants, animals, and, of course, humans. However, no matter what the size of the organism is, all share the living cell as their smallest repeating unit. And although the cell is in itself an extremely highly organized entity, the genetic part of the cell beats everything in the universe in terms of degree of order. The way that nature has developed an administrative system that passes on hereditary information from generation to generation is mind-boggling. Let me illustrate this with a few numbers.

The genetic material in a cell is located in the chromosomes, of which the human cell has 23 pairs. All the properties of the human body, which starts with the fertilization of the female egg cell by the male spermatozoa, are contained in that genetic mass. After that first cell, there are about 50 generations of cell divisions needed to build a complete human body (total number of cells after 50 divisions is about 10^{15} cells)[90]. The fundamental information carrier in DNA is based on combinations of only four molecules: adenosine (A), cytosine (C), guanine (G), and thymine (T). They always show up in pairs (AT and

---

[88] Recall that W stands for the number of realizable microstates that together form the given macrostate.

[89] One can argue that the cell is not the smallest form of life, but rather viruses or prions. However, these cannot form structured organisms by themselves, whereas cells certainly can.

[90] One cell will form 2^{49} cells after 50 divisions, which is about 10^{15} cells.

CG) in the DNA double helix. Within the DNA strand, the 46 human chromosomes contain about 6 billion base pairs (with each chromosome containing several hundred million pairs) and account for about 2 gigabit of information. Each chromosome contains several hundred million base pairs. However, only 2% of the total DNA mass is used for the genes, each of which encodes particular body features, such as eye color. Yet there are about 30,000 genes in the human genome. Is it not unbelievable that this tiny amount of material can store such a massive amount of information that in turn controls height, musculature, hair color, mental abilities, gender, and everything else that makes up a human body? The efficiency of information storage, and its proliferation from cell to cell, can only be called extremely well organized. Nowhere else in the universe (as far as we know) can you find such a high degree of order! But most surprising is that life not only creates this high degree of order, but also maintains it against nature's tendency to increase disorder. We'll come back to this remarkable property of life later in the chapter.

The statistical nature of physical laws; or, to make something happen, atoms and molecules need to work together in large groups

Let's detour a little to elaborate on an observation originally made by Schrödinger. (This section is intended as background information and is not needed to understand the rest of this chapter).

Because atoms and molecules are so small, for natural events to happen, a great many atoms or molecules must be involved. For instance, just one liter of air involves about 10^{22} molecules. Because of these large quantities, physicists have resorted to describing average characteristics of these ensembles of molecules, rather than trying to describe each individual molecule, when describing macroscopic events such as pressure, temperature, density, and our old friend, entropy. They do this by using statistical analysis. Boltzmann used statistical models to describe properties of ensembles of atoms and molecules, which eventually led him to his famous expression for entropy. But Schrödinger was one of the first to point out that physical laws were always statistical in nature. In order to see any macroscopic effect, massive amounts of atoms and molecules need to work together – otherwise, our sensory organs would simply not record any effect. At the same time, it is also true that the larger the number of involved atoms, the better the statistics apply. We have seen that in Brownian movement, each particle follows a path that is truly random. However, in large amounts, these particles can have a net effect when placed in a concentration gradient. Then the net effect is that of diffusion: the particles drift away so that the concentration gradient diminishes, or disappears altogether.

But what do we see in the genetic field? Here we conclude that the genes, which carry the fundamental code for organism features, can consist of relatively small amounts of atoms or molecules, say 500000 atoms on average or so. That is much less than the amount in a liter of air. Thus, we see that in a living organism, the fundamental hereditary process is NOT necessarily following statistical physical laws. Both statistics and entropic analysis work only with large numbers; one cannot talk about the entropy of a single molecule or atom. For example, the Brownian track that is followed by one individual particle does not predict the aggregate movement of the billions of particles around it; but in genetics, the exchange of one molecule by another (e.g., replacing adenosine with cytosine in one DNA location) can have dramatic consequences for the organism's macroscopic characteristics – such as hair color or gender.[91] Plus, the molecules that form genetic building blocks are very stable; under normal conditions, they will not disintegrate, and so stubbornly retain their genetic information.

Life and entropy

We have seen that isolated systems, when left alone, will progress to a state of maximum entropy, thus constantly showing an increase in entropy, with the resulting disorder. An organism, however, represents an extremely high-ordered system. The cell, by a very high degree of chemical engineering driven by its DNA codes, can steer all kinds of chemical reactions and diffusion processes (also called metabolism) to maintain its structure. When we go from the elementary cell to the organs and then to the human body, we must conclude that the entropy of living organisms, be they single cells or human beings, is relatively low. We deduce that the complex chemical machinery that makes life possible has

---

[91] And indeed, this is what happens in mutations. A mutation is simply a local change in the genetic code caused by natural or manmade radiation, or certain chemical actions. These can lead to the formation of some of the most harmful mutations – cancer cells!

a distinctive result: it keeps the entropy of the organism low. This leads to some really interesting ideas.

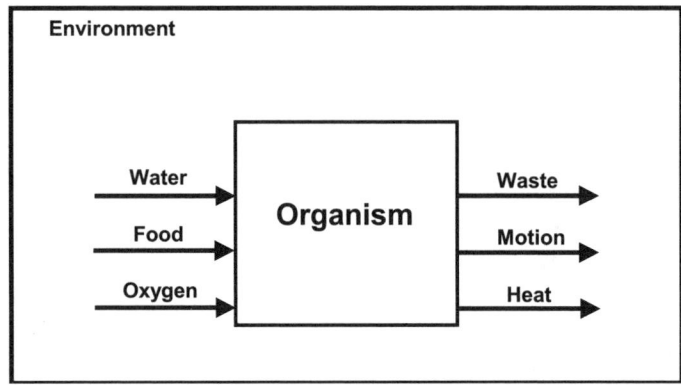

Figure 6.1 Schematic flow diagram of a living organism.

First, it may seem that the ability of living organisms to limit their entropy contradicts the Second Law, which says that isolated systems increase their entropy all the time. But we're overlooking something fundamental here: *living organisms are not isolated from their environment at all*; in fact, they maintain an extensive exchange of materials and energy with their environment! Let's have a closer look at how this works by building a simple model of the situation, as in Figure 6.1. A living organism is involved in a continuous and steady state[92] process of taking up low-entropy food and energy from the environment. In doing so, the organism is able to keep its own entropy at a low constant level, but still produces a high-entropy output in the form of waste and heat. The surrounding environment, and indeed the entire isolated system (defined by the outer box in Figure 6.1) will see an increase in entropy[93].

---

[92] In this case, steady state means that the amount of energy used by the organism is the same as that dissipated into the environment (but of course, with a qualitative change from available energy to the non-available form!).
[93] Wondering how an isolated system can increase its entropy? See the case in Appendix V.

6: Energy, Entropy, Life, and Heat Death   119

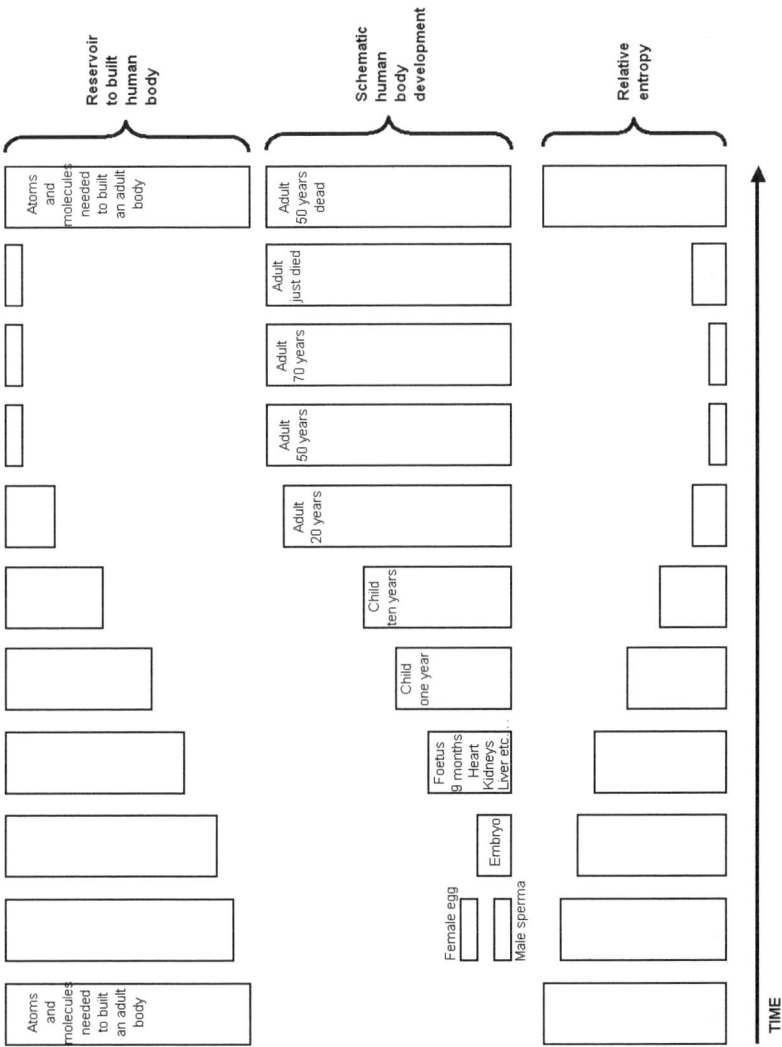

Figure 6.2 Entropy trend during the life of a human.

Second amazing idea: it appears that the ability to limit entropy is characteristic of life. We could say that living organisms differentiate themselves from dead materials or systems because they have found a way to keep their entropy low. Non-living materials, such as rocks, are in a continuous process of decay. Natural processes such as running water and wind continuously erode tiny particles from the rock, eventually rendering it to sand. This is why mountains slowly disappear and why older mountain chains have lost their sharper edges while younger chains are much more defined – and all of it dictated by the Second Law! Schrödinger put it quite nicely: "The organism is sucking orderliness from its environment." Treating the entropy level time trend as a criterion is certainly one way to define life.

When a living organism dies, it starts to decay rapidly. Its structure falls apart and its components are eventually given back to the environment in the form of simple molecules, such as water and iron oxide. It is clear that during this process of decay, the entropy is no longer controlled, but instead increases continuously until it levels off at the state of maximum entropy. Figure 6.2 shows how the entropy of an organism changes during different phases (before conception, growth, adulthood, death, and after death). In the upper region of the figure, we see the stock of atoms and molecules from which the human body is built. (In this case, the stock contains just a bit more than what is actually needed.) Of course, the stock is depleted as the organism grows from conception to embryo, fetus, and so on. If we wait long enough after death, the stock will be refilled with the same atoms and molecules that existed before conception.

In the middle part of the figure, we see schematically how the human body develops over time. In the lower part, we observe how the entropy changes. Before conception, when we have only the reservoir of atoms and molecules, the entropy is relatively high because there is no order in the reservoir. All the atoms and molecules are mixed randomly. However, after conception, when the body starts to build, the entropy decreases because of the tremendous amount of molecular organization going on. To maintain that low level of entropy is literally a matter of life and death. But, after the cessation of the human's life, the entropy will increase and return to its initial value (note that in Figure 6.2 we're only talking about the human body, exclusive of its environment).

Entropy and the food chain.

Entropy and the planet

In Appendix VIII on photovoltaic cells, we see that the earth continuously receives a stream of solar energy averaging about 1400 Watt/m^2. We have also seen in this chapter that life on earth needs a constant source of available energy to maintain a low entropy state in order to exist. In the following text, we will do a step-by-step analysis to investigate how sunlight changes the entropy content of our planet, and the role that living organisms play in this process.

Consider first the situation in Figure 6.3.A, with the sun radiating its light into the dark universe. The sun's surface temperature is about 5500 Kelvin, while the surrounding space is at a temperature of 3 Kelvin, close to absolute zero. Per a given time unit, the sun radiates an amount of energy $-\Delta E$ that is absorbed as an amount $+\Delta E$ into the universe[94]. Since the pressure of the universe is very low (almost a vacuum), there is no work involved in this transaction of energy and the work term $P\Delta V$ is nearly zero and $T\Delta S = \Delta E$. Therefore, it is easy to calculate the entropy changes in this system:

Entropy change for the sun: $\Delta S_s = -\Delta E/5500$ and
Entropy change for the rest of the universe: $\Delta S_{Univ} = +\Delta E/3$

Therefore, the total entropy change of the system is:

$$\Delta S_{Total} = \Delta S_{Univ} + \Delta S_{Sun} = \Delta E(1/3 - 1/5500) \qquad (6.1)$$

and it is not difficult to see that ΔS_{Total} will be a number larger than zero. This should not surprise us. The universe can be considered an isolated system and the sun radiating energy is a spontaneous process. In Chapter 2, we found that for spontaneous processes in isolated systems, the entropy change must always be larger than zero; thus the entropy must always increase.

---

[94] In this part of the book we change the symbol for energy from U to E. In the literature also both letters are used.

Now imagine our planet in a time when there was no life, nor any atmosphere or surface water – in other words, like the moon or Mars today. The sun radiates (per second) an amount of energy of $(-\Delta E)$ into the universe. A smaller proportion of that energy is absorbed by the earth $(+\Delta E_{Earth})$. However, we know that on average, the temperature of the earth is stable and does not change. Therefore, we conclude that in any given time period (say, 24 hours) the earth radiates an equal amount of energy back into space[95],

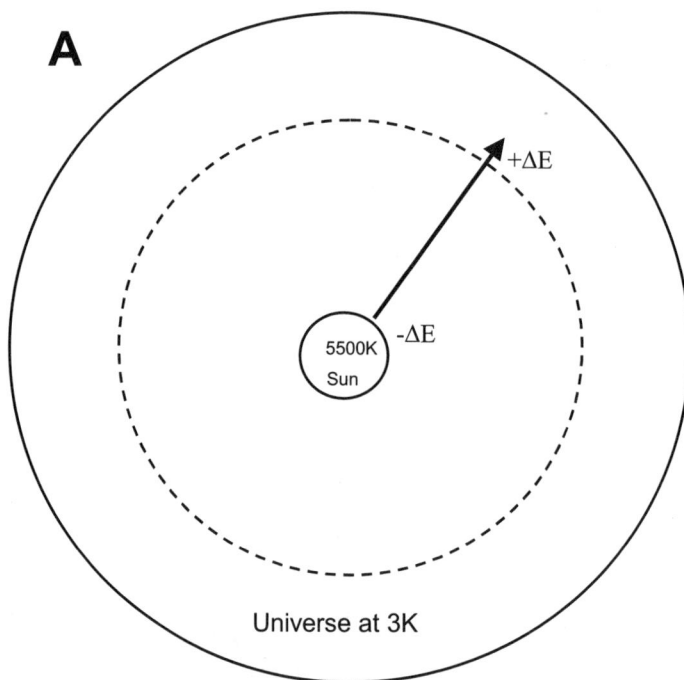

Figure 6.3.A The sun radiates an amount of energy ΔE into the universe.

and thus the total entropy balance is (see Figure 6.3.B) :

---

[95] This situation is similar to the steady state of a living organism: the influx of energy is equal to the outflow of energy.

6: Energy, Entropy, Life, and Heat Death

$$\Delta S_{Total} = (\Delta E - \Delta E_{Earth})/3 - (\Delta E - \Delta E_{Earth})/5500 - \Delta E_{Earth}/5500 +$$
$$+ \Delta E_{Earth}/255 - \Delta E_{Earth}/255 + \Delta E_{Earth}/3 \qquad (6.2)$$

Some simplification reworks the above formula into:

$$\Delta S_{Total} = \Delta E(1/3 - 1/5500) = \Delta S_{Univ} + \Delta S_{Sun,} \qquad (6.3)$$

all of which is represented graphically in Figure 6.3.B below,

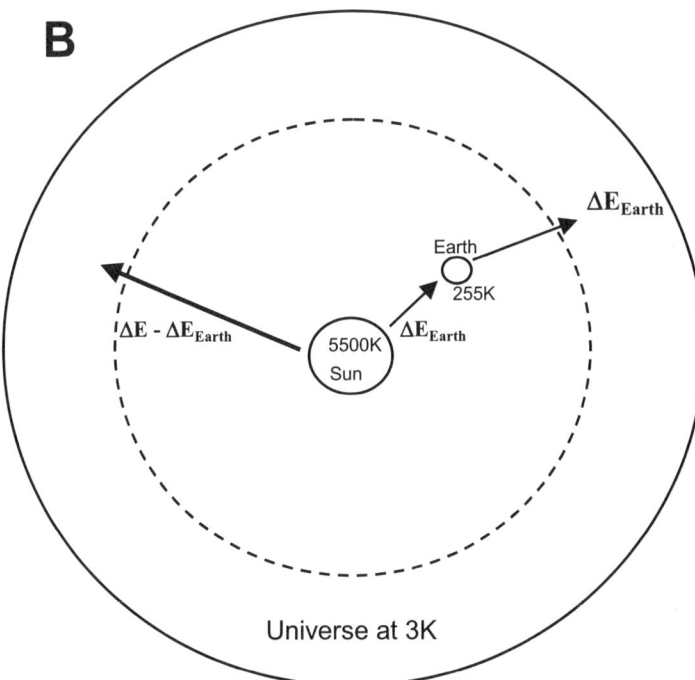

Figure 6.3.B The sun radiates energy to both the earth and the universe, while the earth re-radiates its solar energy into space.

and this result is identical to that shown in Figure 6.3.A. This shouldn't surprise us too much, since the net energy change of the earth is again zero and the earth's entropy change is zero as well[96].

---

[96] However, Appendix V shows that entropy is created in a system by expanding gas while there is NO change in energy.

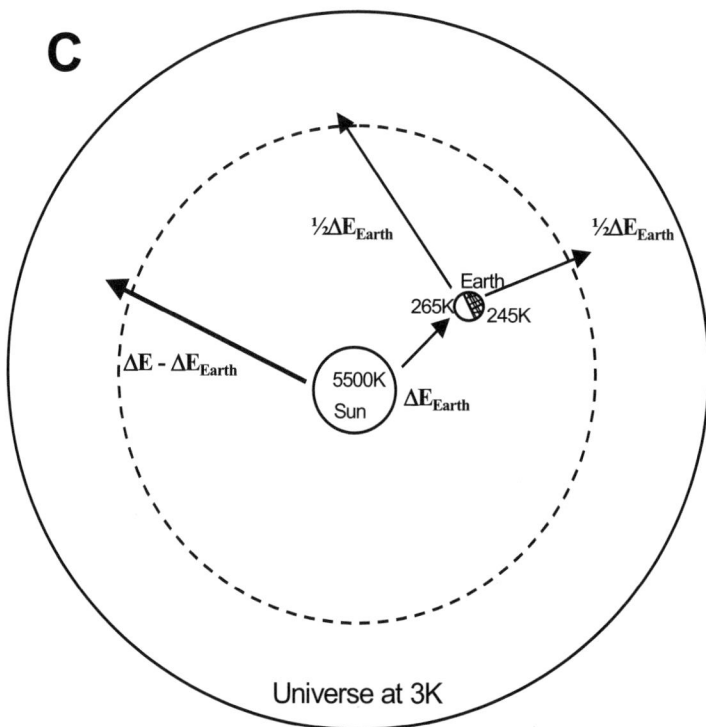

Figure 6.3.C In this situation, the earth now has a cold side (night) and a warm side (day), and shows an overall entropy decrease (see text for more details).

However, the situation in 6.3.B is an oversimplification of reality, and so we have to refine our model. A more realistic assessment is that at any point, sunlight illuminates only about one half the planet,[97] while the other half, of course, experiences nighttime. Earth's dayside has a somewhat higher temperature than the night side (see also Figure 6.3.C). Let's assume the average temperature difference is 20K, with the day hemisphere at 265K and that the night hemisphere at 245K. As we will see, this changes the entropy balance. Although the energy flux from the sun doesn't change, the absorption of energy by earth now happens at

---

[97] The input energy flux of sunlight is about 1400 Watts /m^2 of which about 30% is reflected back into space by clouds, ice, snow and dust particles.

6: Energy, Entropy, Life, and Heat Death

265K ($\Delta S = +\Delta E_{Earth}/265$) while the earth's radiation of solar energy occurs in a gradually declining range between 265K and 245K. However, let's assume that $1/2\Delta E_{Earth}$ is radiated at 265K and that $1/2\Delta E$ is radiated at 245K. Thus, the total loss of energy by radiation is still equal to the total gain by absorption, a balance needed to prevent our planet getting too hot or too cold. The entropy balance is then:

$$\Delta S_{Total} = (\Delta E - \Delta E_{Earth})/3 - (\Delta E - \Delta E_{Earth})/5500 - \Delta E_{Earth}/5500$$

$$+ \Delta E_{Earth}/265 - \frac{1}{2}(\Delta E_{Earth}/265) - \frac{1}{2}(\Delta E_{Earth}/245) + \Delta E_{Earth}/3$$

or more simply:

$$\Delta S_{Total} = \Delta E(1/3 - 1/5500) + \frac{1}{2}(\Delta E_{Earth}/265) - \frac{1}{2}(\Delta E_{Earth}/245) \quad (6.4)$$

The last two terms in expression 6.4 are, in fact, the entropy change for the earth and can be written as:

$$\Delta S_{Earth} = \frac{1}{2}\Delta E_{Earth}(1/265 - 1/245)$$

However, we still overlooked one fact. That is that although we radiate $\frac{1}{2}\Delta E_{Earth}$ to space at 265K and at 245K we *received* this amount of energy from the sun at 265K! But we know now that if we transport energy from a reservoir at 265K to a reservoir at 245K, the associated entropy change is:

$$\Delta S_{265 \to 245} = \frac{1}{2}\Delta E_{Earth}/245 - \frac{1}{2}\Delta E_{Earth}/265$$

and thus the total entropy of the Earth as a result of the captured radiation from the sun is exactly:

$$\Delta S_{Earth} + \Delta S_{265 \to 245} = 0$$

We conclude that there is no entropy change for a dead planet that is just a piece of rock. It must be said that the assumptions above are still rather crude,[98] but good enough to illustrate this fundamental point.

However, the system depicted above is still too simple to really capture closely what's happening on earth. Our planet is not just a piece of rock, but has a dense atmosphere, oceans, and polar caps. The planet not only has more temperature zones than just between dayside and night side but also between the warm equator and the cold poles. Furthermore, we have energy fluxes present in the atmosphere and in the oceans to balance the temperature gradients. Let's look at it more closely by incorporating some of these energy fluxes.

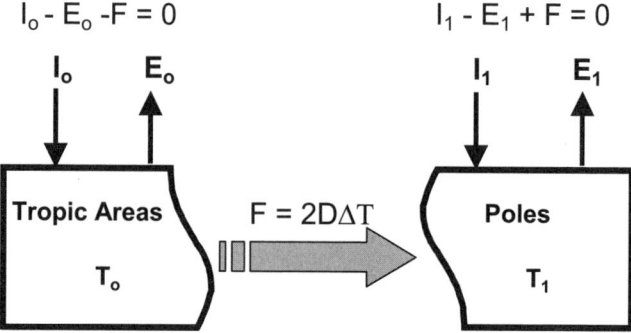

Figure 6.4 Simplified model of the energy flows present in the earth-atmosphere system (after Lorenz et. al., 2001).

In the situation of Figure 6.3.C above, we incorporate a two-zone temperature difference on earth, day and night. And although the temperature difference caused by day and night is not permanent for a given location on the surface, it is averaged on a 24-hour- time scale. A bigger and permanent temperature difference is present between the tropical and polar latitudes. Heat transport through the atmosphere tends to make that temperature difference smaller[99]. Geophysicists have used

---

[98] The crudeness lies in the assumptions for the temperature difference and the distribution of that difference. Also, the amounts of radiation received and transmitted are a function of temperature as well.

[99] Actually, heat transport also occurs within the earth's crust, but we won't consider that here.

6: Energy, Entropy, Life, and Heat Death

the concept of entropy to predict the temperature difference between the tropical and polar regions not only for the earth, but also for other heavenly bodies such as Titan and Mars[100]. Below we will describe the different heat flows with a model developed by researchers from the University of Arizona and the NASA Ames Research Center [Lorenz et. al. 2001] (See Figure 6.4).

The earth absorbs an amount of energy I_0 in the form of heat in the tropic region, and likewise an amount of I_1 in the pole areas. One reason that these amounts are not the same is because of the different textures in these areas (soil and plants in the tropic area, and ice in the polar areas) lead to different reflective properties. In addition, the different incident angles have an influence as well. At the same time, while receiving or absorbing energy, these areas radiate an amount of energy back, E_0 and E_1, into the cold universe — in the same way as a black body, which we discussed in Chapter 4. All this results in a temperature difference between the tropic and polar regions. The atmosphere, however, is able to transport heat by diffusion or by convection (with winds). This lessens the temperature difference between the poles and the tropics. The atmosphere's capacity to conduct heat can be expressed in a number called the *meridional heat diffusion coefficient*, or D. For earth, the value of D has been empirically calculated as between 0.6 and 1.1 W/m²K and the observed heat flow, F, has been reckoned between 20-40 W/m². The heat flow depends on D and the size of the temperature difference (ΔT), and can be expressed as F = 2DxΔT (the factor of 2 represents the twin flows from the equator to the north pole and south poles). The heat transport from the tropics to the polar regions is similar to the flow of heat from a warm location to a cold reservoir (see Figure 6.4). In that case, the entropy production ($\Delta S/\Delta t$) is simply[101]:

$$\frac{\Delta S}{\Delta t} = F(\frac{1}{T_1} - \frac{1}{T_0}) \quad (6.5)$$

---

[100] What causes the temperature difference between the tropical and polar regions? One reason is that the polar regions absorb much less heat because their icy surfaces reflect sunlight more efficiently.
[101] Note that heat flow is the amount of heat per second. That's why the entropy increase (based on heat flow) is also measured per second, and so entropy production is expressed as $\Delta S/\Delta t$ in equation 6.5.

From this simple equation we can draw a few conclusions. If D is very small, then there will be no heat transport ($F = 0$), and therefore no entropy production. This means that the atmosphere does not conduct heat and the temperature difference between the poles and the equator is maximal (left side of Figure 6.5). If D is very large, we have maximum heat transport, but this is now so effective that T_0 and T_1 will approach the same value and again, the entropy production will be very small. In this case, the temperature difference between the poles and the equator will be minimal (right side of Figure 6.5). We intuitively feel, however, that there will be an intermediate value of D that will lead to a situation of real entropy production. Thus, if we plotted the entropy production versus D, we would expect that $\Delta S/\Delta t$ would attain maximum value. It has been shown that indeed, the atmospheric system of the earth does reflect this situation, called the Maximum Entropy Principle, or MEP. This principle seems to have a general validity in systems that a) receive a constant flux of energy from outside the system and b) that are far away from equilibrium but in a steady state (this is indeed the case for our planet).

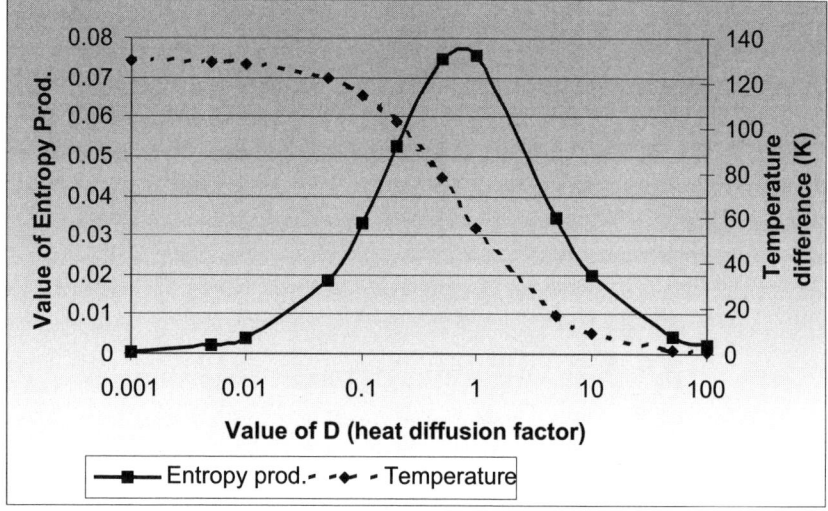

Figure 6.5: The temperature difference between the tropic and the polar regions, and the value of the corresponding entropy production as a function of the meridional heat diffusion (D).

6: Energy, Entropy, Life, and Heat Death

Figure 6.5 uses data from Lorenz to plot the entropy production and temperature difference versus the heat diffusion coefficient, D. It is clear that the entropy indeed reaches a maximum for a D value of about 1 W/m^2K, reflecting a temperature difference of about 55K. This is a credible result (considering the value of 1 for D in the empirical range mentioned above), and the temperature difference of 55K is certainly a reasonable value, too. Certainly, the real situation on earth is much more complex than the model suggested in Figure 6.4[102]. However, the same maximum entropy principle has been used for Titan and Mars, and in both cases, with remarkably good results for the prediction of temperature differences.

So, the overall conclusion is that for a planet with an atmosphere (made of gases such as nitrogen, oxygen, water etc.), the energy radiation of the sun leads to an *increase* of entropy of that planet. The production for earth has been estimated to be about $893 mJ/m^2Ks$ or about one joule per square meter per second and per degree Kelvin [Kleidon and Lorenz, 2004].

But what happens when we have a planet where life forms are present? Let's assume that this planet is covered with extensive forests and that part of the incoming solar energy is used in photosynthesis. It can be shown that theoretically the maximum efficiency of photosynthesis is about 4.5% [Walker, 1993]. However, many unfavorable factors such as water shortage or limiting CO_2 concentration bring the actual photosynthetic yield down. Less than 1% of the available solar energy is actually transformed into chemical energy to sustain plant life and will therefore not impact the entropy balance of the planet substantially. In the next sections we will highlight photosynthesis in more detail.

Energy and Entropy of the food chain

All life on earth is made possible by the light of the sun. We have seen in the previous sections what the associated entropy and energy flows are. We also have seen that the entire redistribution of energy from the sun to the earth and the rest of the universe comes with an entropy increase, exactly as the Second Law predicts. However, we also have seen that life

---

[102] For instance, there is also heat transport in the oceans and in the earth's crust, which also contributes to the production of entropy.

needs a low entropy resource - photosynthesis - to survive and reproduce. Used by plants, algae, and some bacteria, photosynthesis involves a photochemical reaction that leads eventually to a process called carbon dioxide (CO_2) fixation. But before we discuss this elegant process, let's first have a look at the food chain here on earth (see Figure 6.6).

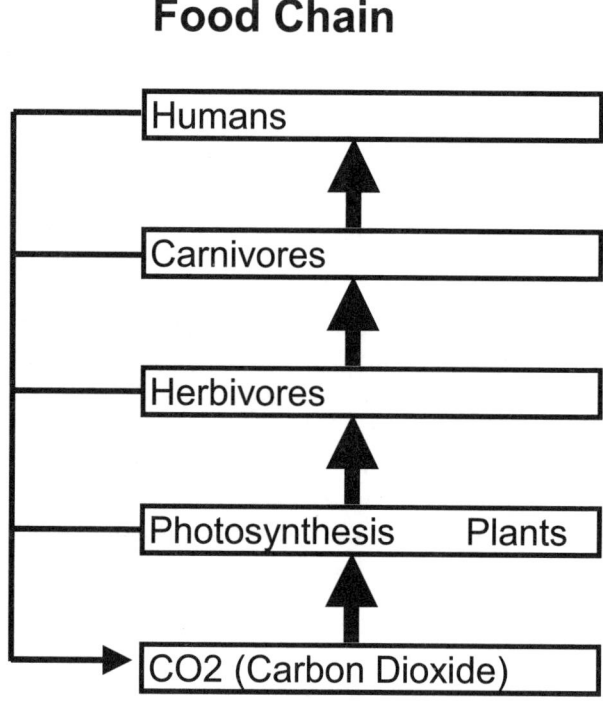

Figure 6.6 Simplified food chain. In each upward arrow the efficiency of the chain is only 10%. This means, for example, that humans can only use 10% of the available energy from the food they eat.

The principle is well known: each step in the food chain serves as nutrition for the creatures on the next step, and each animal produces carbon dioxide while consuming oxygen from the atmosphere. What is less well known is that there is a large amount of inefficiency in this food chain. (Efficiency is defined here as the proportion of energy actually used by an organism, compared to the total energy present in its food.) This can be illustrated with several examples [Glencoe, 2004]: 100 kilograms of grain are needed to produce 10 kilograms of beef, which create only 1 kilogram of human tissue. Similarly, 3000 blades of grass are needed to produce 250 grasshoppers, which will feed 25 birds that will be eaten by just one fox. In general, each higher level in the food chain transforms only 10% of the energy from the next level beneath it. From that point of view, it is rather inefficient to feed cattle with grain, and then eat the cattle; ecologically, we would be much better off if we just ate the grain ourselves[103]. True, energy efficiency isn't very high in the photosynthetic process either (as we will see shortly), but the difference here is that solar energy comes in such abundant quantities that efficiency is not a concern!

So, all life on earth starts with sunlight and the photosynthetic process. Simply put, light quanta from the sun are used to drive chemical reactions that lead to the formation of oxygen and the synthesis of carbohydrates. The overall chemical process can be written as:

$$6H_2O + 6CO_2 + h\nu \rightarrow C_6H_{12}O_6 + 6O_2 \qquad (6.6)$$

In ordinary language, this says that six molecules of water and six molecules of carbon dioxide are transformed into one molecule of sugar and six molecules of oxygen. (The term $h\nu$ stands for the light quanta that are needed to drive the reaction.) Massive research has shown that the fundamental chemical reactions involved in producing sugar and oxygen are the same in all photosynthetic organisms. The structure of a common sugar, β-D-glucose, is as follows:

---

[103] A few examples to illustrate the inefficiency of the "artificial food chain" [from Manning, 2004]: The agriculture industry needs about 35 J of fossil fuel to produce one J of beef and about 68 J to produce one J of pork. Processed food requires about 10 J to produce one J of food energy.

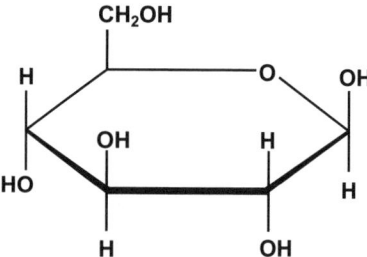

Although the overall reaction in equation (6.6) suggests a rather simple mechanism, the reality is that photosynthesis is extremely complex, and even today is not completely understood. The first step at unraveling the process was Joseph Priestly's discovery in 1770 that leafy plants produce a gas (oxygen) that supports combustion. In 1845, our old friend, Julius Robert von Mayer, conjectured that plants convert light energy into chemical energy.

Just a few numbers will illustrate this essential life-enabling process [Whitmarsh, 1995]:

- Producing 1 oxygen molecule requires about 8 (red[104]) light quanta. To make 180 grams of glucose there is needed about 3000 kJ of energy.
- Each year, 10^{14} kilograms of carbon are removed from the atmosphere by photosynthesis. The energy needed to do this is only 0.1% of all the solar energy captured by earth.
- Every year, more than 10% of the total atmospheric carbon dioxide is converted into carbohydrates (or glucose, a 6-carbon sugar).

The pigment that absorbs the light is called chlorophyll. Typically, only visible light is absorbed, predominantly in the red and blue wavelengths, and less so in the green wavelength. That absorption pattern gives chlorophyll-laden plants their green color.

Photosynthesis can be divided in two major steps: the oxygen producing step (photophosphorylation) and the carbon fixation step that

---

[104] The color of the light quanta is significant, since quanta energy depends on their frequency, each of which has a characteristic color.

eventually produces glucose (also called the Calvin-Benson cycle) – see Figure 6.7.

The first step, photophosphorylation, is a light-enabled reaction in which water is consumed and oxygen and molecules of ATP and NADPH are produced. ATP stands for adenosine triphospate and NADPH stands for nicotinamide adenine dinucleotide phosphate. To this day, we don't completely understand all the steps and chemicals involved in making this happen. That said, we'll limit ourselves to a brief description of how ATP and NADPH carry energy. ATP can store within its chemical bonds a large amount of solar energy. Each molecule of NADPH can carry excited electrons, which is another way of storing energy. These two molecules are the primary energy carriers in the living plant cells.

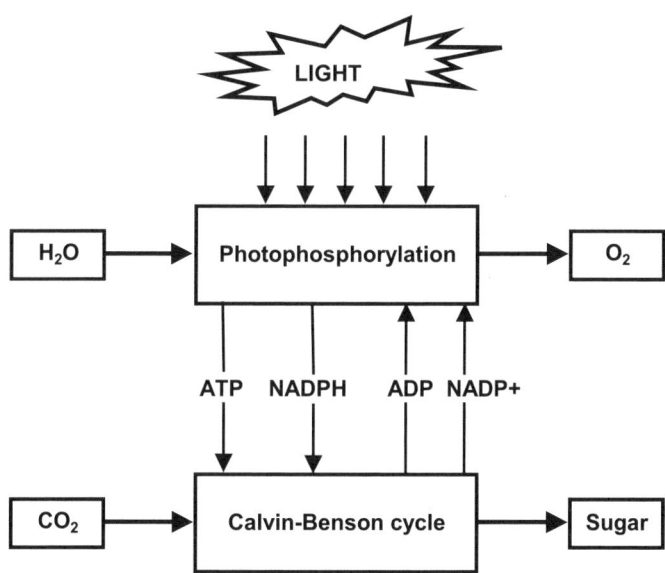

Figure 6.7 A very schematic and simplified representation of the photosynthetic process. ATP, ADP, NADPH, and NADP+ are energy carriers. While photophosphorylation needs light, the Calvin-Benson cycle can run in the dark.

The second stage (or Calvin-Benson cycle) can work without light. Using the energy of the ATP and NADPH molecules and carbon dioxide from the atmosphere, this cycle creates chemical reactions in the cell that eventually form sugars (glucose).

How about the entropy change for the reaction in equation (6.6)? On the left side are six water molecules and six carbon dioxide molecules. Although both sets of molecules are in liquid state during the reaction within the cell, eventually they are released as gases to the atmosphere. Altogether, they are less complex than the single sugar molecule and six oxygen molecules on the right side. Therefore, we expect the entropy to decrease during photosynthesis, which is driven by the energy in sunlight. Indeed, quantitative calculations[105] show that the entropy change for the overall reaction in equation (6.6) is negative and can be calculated as ΔS = -976 J/K (for one mole[106] of glucose, which is 180 grams). Because the entropy decreases, this reaction cannot proceed spontaneously and therefore must be driven by an external energy source which is, of course, sunlight. Thus, photosynthesis counteracts the entropy production caused by human actions — but we must not forget that only about 0.1% of total sunlight drives this entropy decrease. Therefore, we expect that the entropy balance of the earth as a whole will not be influenced substantially by photosynthesis.

The opposite of reaction (6.6) will occur when the organism needs energy. This process is called the respiration of glucose, and is in fact the combustion of glucose under well-controlled circumstances in the cell. The products of that reaction are water and carbon dioxide:

$$C_6H_{12}O_6 + 6O_2 \rightarrow 6H_2O + 6CO_2 \qquad (6.7)$$

How much energy will that reaction deliver? A lot: for example, 180 grams of glucose (about 40 sugar cubes) will generate almost 3000 kJ of energy – enough to allow a 75kg human to climb a mountain of about 4000m high! Impressive isn't it? At the molecular level, reaction (6.7) produces energy to "charge" molecules of ATP and NADHP. In the case of ATP, this happens by adding a phosphate group (PO_4^-) to ADP, which

---

[105] The calculation is rather simple, thanks to scientific tables that provide the standard entropy for many chemical compounds. Therefore, the standard entropy (in thousands of J per mole) is 189 for water, 214 for carbon dioxide, 205 for oxygen, and 212 for glucose. Thus, the total entropy change in equation 6.6 works out as: ΔS = (products) − (reactants) = (212 + 6x205) − (6x214 + 6x189) = -976 J/K mole.

[106] One mole is a chemical unit of quantity and is $6.02 \cdot 10^{23}$ atoms or molecules.

means that for every molecule of glucose, two molecules of ATP and two molecules of NADHP will be "charged."

A sort of artificial "photosynthesis" is also possible with photovoltaic, or solar cells. These devices use a rather elegant process to convert sunlight into electrical energy. The electrical current generated by the solar cell can then be used to split water in hydrogen and oxygen, and the resulting gases can be used in fuel cells to produce electrical power and water. On paper, this looks like a very attractive energy conversion technology, without the pollution of nuclear and fossil fuel based power plants. Appendix VIII elaborates on the function and energy balance of solar cell technology.

In summary, we have seen how plants, using complex photochemical reactions, convert energy from the sun into chemical energetic "storage" molecules. This process enables almost[107] all life on our planet, both by supporting the food chain and by maintaining the atmosphere's delicate balance between oxygen and carbon dioxide.

Heat Death

All objects with a temperature above absolute zero can be considered "black" bodies. That is, they all radiate electromagnetic radiation (light) to the cooler environment, i.e., cold space. We have already seen what this means for the sun and earth in terms of entropy development: the universe will continue to evenly distribute the energy, and therefore increase its entropy constantly.

In 1856, Helmholtz deduced the implications of the Second Law at the universal level. He said the universe would drift toward a state of maximum entropy, until all available energy reached the latent or unavailable form. Put another way, all energy will have been distributed evenly, and temperature differences will no longer exist. We know from Chapter 2 that it is impossible to convert energy into work if there are no warm and cold heat reservoirs! And so, the universe will come to a state of complete rest and will no longer change at the macroscopic level. Helmholtz named this situation Heat Death, which also can be called thermal equilibrium for the universe.

---

[107] Almost, because there are some forms such as the chemosynthetic organisms near ocean volcanic vents that live without relying on photosynthesis

An interesting question in this respect is why, 10 to 20 billion years after the Big Bang, all the primordial energy is still not evenly distributed throughout the universe. What's taking the universe so long to reach thermal equilibrium? This question has been addressed by many scientists. One such thinker, Paul Davies [Davies, 1972], came up with three basic answers. First, if the universe had reached thermal equilibrium, we would simply not exist and therefore wouldn't know the difference. That begs the question, though, and so Davies' second answer is that the universe, having begun in a low entropy state, is on its way to equilibrium and is simply taking a long time to get there. But this answer doesn't tell us much. The third answer, more sophisticated, notes that we're dealing with an expanding universe. Because of this expansion, the universe is not a static system but changes all the time. By constantly expanding, the universe behaves as a tremendous heat sink that continuously redistributes heat, with an associated entropy increase. This last answer gives us a better feeling for why the universe is still not at equilibrium, and that we have our dynamic cosmos to thank for it.

7
The Use of the Concept of Entropy in Other Sciences

In Chapters 5 and 6, we saw how the concept of entropy found its way into fields other than physics, influencing economics, environmental science, and biology. In this chapter, we will look at other fields even more distant. Although importing entropy into these fields has sometimes produced very usable results and new insights, it also has led to confusion in many cases. In a way, one can say that it all started with the order and disorder interpretation of Boltzmann, followed by the work of Claude Shannon, who used entropy in his famous treatment of communication theory in 1948. Shannon derived a formula describing the density of information that could be transported through a wire, a computation that very much resembled the formula that Ludwig Boltzmann derived for entropy (see Chapter 3). Shannon thus named his expression the "entropy of information." Although there are some similarities between the entropies of Boltzmann and Shannon, from a physical point of view there are fundamental differences, and one can question whether Shannon's use of the word *entropy* was an appropriate choice. However, the statistical abstractions made by both Boltzmann and Shannon, and the subsequent popularization of the concept, led sometimes to bizarre discussions. The genie was out of the bottle, and since then the concept of entropy has been used in many other fields, which we will discuss shortly.

Entropy and electrical communication

A brief history of electrical communication

The history[108] of modern communication in the US really starts with the discovery by Samuel Morse in 1835 that electrical current could be used to transmit signals through a wire. On May 23, 1844, the first telegraph line between Washington and Baltimore was officially opened and bridged a distance of 40 miles. In 1851, Western Union established the first transcontinental telegraph line. At that time, one could transmit about 50 words per minute. Later developments, such as *multiplexing*, took the transmission capacity up to several thousand words per minute. In 1876, a rival to long distance telegraphy emerged: the telephone. Alexander Graham Bell was able to transmit for the first time a verbal statement over a wire to his co-worker, Thomas Watson: "Mr. Watson, come here, I want you." In 1876, Bell was awarded a patent and two years later Bell started a telephone company that eventually became American Telephone and Telegraph (AT&T). By 1880, there were almost 50,000 phones in the United States. Transmission capacity improved continuously and by 1936, a pair of coaxial cables could carry nearly 2000 telephone conversations. Finally, in 1956, the first trans-Atlantic submarine telephone cable was established between Newfoundland and England. Today, telephone communication between the continents is still done with submarine cables but now the copper wires are replaced by fiber optic transmission technology.

Communications now reached into the wireless realm. In 1886, Heinrich Hertz verified experimentally the existence of radio waves, which had been predicted in 1860 by James Maxwell. This was a huge triumph of science. However, it was Marconi who translated these scientific results into a workable radio transmitter and receiver. Around 1900, ships were among the first to take advantage of this new mode of communication technology. And in 1910, Marconi established a regular American-European radio communication service. Later years would bring other developments, such as transmitting pictures through a wire, TELEX, facsimile, cable television, satellite communications, and Internet communication through the World Wide Web. Transmission

---

[108] A nice overview can be found on the Federal Communications Commission (FCC) website at www.fcc.gov/cgb/evol.html.

speeds are still going up, although their theoretical limits appear to be one of the big roadblocks to continuously faster wireless and Internet communication. In the next section, we will learn a little more about the "science" behind communication technology.

But let's do some tourism first. From 2002 until 2005, I was on assignment for my company in Austin, Texas. My family and I took advantage of our stay in the US by visiting historical sites. In 2005, we took a short vacation to Boston, MA to witness the Fourth of July celebrations, including the proclamation of the Declaration of Independence from the same balcony where it was read in 1776. Afterward, we traveled to Cape Cod, a popular vacation spot and home to Provincetown, where the Pilgrims of the Mayflower landed. But few people know that it was at Cape Cod where two important milestones were achieved in communication history, with long-lasting impacts to electronic communication through both cable and wireless media.

The first event took place in the French Cable Station at Orleans in 1889, when the first transatlantic, two-way cable between France and the United States, called "Le Direct," began transmitting messages[109]. This system operated as a telegraph, which created signals by subjecting the cable wires to alternating positive and negative voltage, with negative voltage representing a dot and positive voltage a dash, forming letters by Morse Code[110]. The speed of transmission at that point in time was about 25 words per minute. However, the cables were quite vulnerable and often were broken by earthquakes or ship anchors. Today, you can visit the French Cable Station Museum and see a great deal of the original equipment [111].

---

[109] The first transatlantic cable, which could transmit messages within minutes rather than the weeks required for ship-borne communication, was completed in 1858 between Ireland and Newfoundland. This milestone was extensively celebrated in New York City upon arrival of the cable. Unfortunately, the cable broke after only four weeks. New cables were laid out in 1865 and 1866, only to fail again. On July 27, 1866, communication finally was restored through a cable that ran between Valencia in Ireland and Heart's Content, Trinity Bay, Newfoundland.
[110] A trans-Atlantic cable length is about 2500 km. The total resistance is about 2800 Ohms. The voltage applied was about 60 Volts; thus the current to be detected is about 20 mA.
[111] This is a very interesting museum. One is quickly impressed with the unexpected sophistication of the equipment, given the simple positive-negative current signal pattern that was used.

A second, and perhaps even more important event, took place at a location not far from the French Cable Station on a beach called Marconi Station. It was at this site in 1901 that Marconi established the first wireless, two-way transmission of a message from the president of the US, Theodore Roosevelt, to King Edward VII at Poldhu Station in Cornwall, England, which was followed by a reply from the king to the president. The electromagnetic waves were transmitted from a complex network of wires carried by four wooden towers anchored in the beach sand by concrete blocks. Almost nobody was convinced of the need for this new technology at that time, but we know now what an earth-changing historical event that was. Marconi was able to make his invention a success because of his relentless drive to convince people of its potential.[112] Today, there is almost nothing left of Marconi's station but a model showing how the station looked in 1901.

Claude Shannon, the "inventor" of modern electronic communication network theory

In 1948, Claude Shannon, a scientist at the famous Bell Labs, published a paper entitled "A Mathematical Theory of Communication" [Shannon 1948]. This paper is generally considered the birth of modern communication science. Shannon focused on two main themes, the impact of noise on the capacity of the transmission channel, and how to take maximum advantage of the statistical nature of many data sources. At that time, Bell Labs was owned by AT&T, and was one of the best big, industrial laboratories of its day, and home to many great inventions, including the transistor[113] [114]. At that time, much was known about how

---

[112] Guglielmo Marconi was born in 1874 in Bologna in Italy as the son of wealthy parents. He was very impressed with the results of Heinrich Hertz who showed the existence of electromagnetic waves and that these waves could travel wireless over a certain distance. Marconi was able by persistent experimentation to take the distance up from a few meters between the transmitter and the receiver to thousands of miles across the ocean in 1901. He filed in 1896 for a patent for the principle of wireless transmission of messages. Later, in 1900 he also filed a patent in which the frequency tuning of radio transmitters and receivers was claimed.

[113] The transistor was invented at Bell Labs by William Shockley, John Bardeen, and Walter Brattain in 1947, and since then has had an enormous impact on everyday life.

7: The Use of the Concept of Entropy in Other Sciences 141

to transport a signal from one location to another, whether through a wire or airwaves, but the telephone, TELEX, and facsimile had already become standard business tools. In addition, television was becoming a part of the lifestyle of ordinary people.

However, these advances were nothing compared to what was coming. For example, when Shannon wrote his paper, the transfer rate through a cable could support about 1800 telephone conversations simultaneously; today, an advanced fiber optics cable can transmit about 6.4 million simultaneous conversations, and that's not even the limit[115]. Although it's indisputable that such progress would not have been possible without the transistor (and its corollary, the integrated circuit), neither would it have occurred without Shannon's concepts[116]. Shannon contributed in many ways to information and communication theory, but we'll limit our discussion here to the topics that relate directly to Shannon's view of entropy.[117]

First, we need to understand the difference between information and communication. Information can be defined as valuable knowledge

---

[114] Bell Labs did a lot of work to develop the vacuum tube, invented by Lee de Forest in 1906, as an amplifier for long-distance telephone calls. Around 1900, the maximum distance that could be bridged by metal wire telephone communication was about 1500 km—not enough for transcontinental telephone lines (5500 km). Beyond 3000 km, nothing was left of the signal. Aiming to remain a leader in telephone communication, AT&T assigned the problem to Bell Labs, which developed the signal repeater. In 1914, the first transcontinental phone line — containing three repeaters based on vacuum tube amplifiers — was inaugurated between San Francisco and New York (For more details see Riordan and Hoddeson, 1997).

[115] Source: web site of Lucent Technologies

[116] Claude Shannon was born on April 30, 1916. He studied electrical engineering and mathematics and early on was interested in Boolean algebra. As a student, he worked on different topics, such as switching theory. In 1941, in the midst of World War II, he worked at Bell Labs on cryptography. Over a period of eight years, he developed his theory of communication that aimed to find the most efficient source-receiver system possible, finally bringing it to light in 1948. In later years, Shannon worked in many different areas, designing a chess computer, a mechanical mouse named Theseus that could solve mazes, and a strategy for winning at roulette. Shannon received many tokens of appreciation, including the National Medal of Science in 1966, America's highest scientific medal. See also [Gallagher 2001].

[117] To prevent confusion with the entropy of physics, we'll call the entropy in Shannon's information theory "Shannon entropy."

that can be stored in media such as books, sheet music, photos, or videotapes. Once we have stored information, we can transport it through a channel, which in the electronic realm amounts to a cable or a wireless transmitter and receiver. This transport of information can be called communication.

Shannon developed a simple mathematical description for information and the way it can be communicated. He began by asking what parameter would describe the information content of a message in a page of English text, for example. At the time, the "bit" had been introduced as a binary way to store and transmit information; indeed, an early version of it was the use of positive and negative voltages to transmit coded signals across transatlantic cables. Shannon's formula for information content was[118]:

$$S = -K\sum_{i=1}^{N} p_i \log_2 p_i \qquad (7.1)$$

Here, p_i is the probability or frequency that a given character, i, will occur in a given set (for instance, in a page of text)[119]. Shannon called S the entropy of information, and we will talk in more detail about that value below. K is a constant to provide the correct dimensions for S, and need not concern us (here we will equate K with 1). Before we go further, let's illustrate equation (7.1) with an example, based on Table 7.1. Imagine we have a piece of text constructed of only two characters, A and B (obviously not a Nobel Prize-winning piece of literature, but sufficient for our current needs). There are 11 different cases in the table, all with different occurrence frequencies for A and B. In the first case, A = 99% and B = 1%. To complete equation (7.1) for this case, $N = 2$ because we have only two characters, while p1 = 0.99 and p_2 = 0.01. Therefore, the entropy S for this particular piece of text is calculated as:

$$S = -[0.99\log_2 0.99 + 0.01\log_2 0.01] = 0.08 \text{ bits}$$

In the same way, we can calculate the entropy for the second example, which contains 90% As and 10% Bs in a string, as $S = 0.47$ bits. Now,

---

[118] Because we are dealing with binary coding, we'll use a logarithmic function based on 2. Thus $\log_2 x = y$ means that $2^y = x$, or $\log_2 x = 3.32 \log_{10} x$.

[119] The Greek symbol Σ stands for a summation. For example for a three character case, equation 7.1 becomes: $S = 3.32(p_1 \log p_1 + p_2 \log p_2 + p_3 \log p_3)$.

allow me to make a few observations and discover the secret enclosed in these results.

Imagine a string of characters composed of only As and Bs. If we make sure the string contains 99% As and 1% Bs, then the calculated entropy tells us that at minimum,[120] you will need 0.08 bits per character to encode the string in a binary code, which we can then use to transmit through a wire or store in a memory. Thus, if we have a string of 100 characters (with 99 As and 1 B), we will need a minimum of 8 bits to encode the entire string. The same applies if the string consists of 90% As and 10% Bs: the 100 characters require a minimum 47 bits to encode.

Another important feature is that the Shannon entropy does not depend on the length of the text, nor on the informational content. If A appears at 75% and B at 25% (as in the fourth example in the table), and the Shannon entropy is 0.81, the number of bits to encode strings such as AAAB or AABA or ABAA is exactly the same: 4 x 0.81 = 3.24 bits.

We see that the Shannon entropy depends strongly on the frequencies that the letters appear in the text. Shannon entropy is at maximum for strings that have 50% As and 50% Bs; consequently, this is also the case where we need the maximum number of bits to encode, namely one bit per letter. How can we infer that? In the case of 99% As and 1% Bs, we know there is a good chance the next letter will be an A. So from an informational point of view it is a boring case, since not much information is present in the string. However, in the 50% case, we cannot be certain what the next letter in a string will be, since it is equally likely to be an A or a B. And indeed, we could code such a string so that two consecutive As mean yes, two consecutive Bs mean no, and so on. This is reflected in the Shannon entropy; we need more bits to store or transmit such a string, because it contains more information. The Shannon entropy says nothing about the quality of the information; the message can be complete nonsense. All right, you say, but what can I do with all this? Well, hold on – we're about to broach the real topic, namely data compression.

---

[120] Please bear with me; I will define "minimum" shortly.

Table 7.1 Shannon entropy for different letter probabilities

Probability of letter A (%)	Probability of letter B (%)	Shannon Entropy (bits/letter)
0.99	0.01	0.08
0.90	0.10	0.47
0.85	0.15	0.61
0.75	0.25	0.81
0.65	0.35	0.93
0.50	0.50	1.00
0.35	0.65	0.93
0.25	0.75	0.81
0.15	0.85	0.61
0.10	0.90	0.47
0.01	0.99	0.08

Suppose you have a gas bottle that you want to fill with air so you can go scuba diving. The simplest way to fill the bottle is to open it up, let in some air, and then close the valve. But when you take it under water, you notice the air is gone after a few breaths. What to do? Well, you decide to use a pump to squeeze more air into the bottle. That way, your trip under the waves will last much longer.

You can do the same sort of compression when you want to store data (say, a string of text) or transmit data. The capacity of your memory or your transmission line is limited (although technology improvements have really bumped up these limits – more on this later). But, you can use a data compression "pump." Here's how it works: First take a look at Table 7.2. There are seven different strings of text, each composed of only six letters: A, B, C, D, E, and F. The length of each string is frozen at 60 letters. We let the frequency of the letters in the text vary, as indicated in the table. So in the first case, the string could look like:

AAAAAAAAABAAAAAAAAACAAAAAAAAADAAAAAAAAAE AAAAAAAAAFAAAAAAAAA

but it could also look like:

7: The Use of the Concept of Entropy in Other Sciences

AAAAAAAAAAAAAAAAAAAAAAAAAAAAAAAAAAAAAAAAAAA
AAAAAAAAAAAAAAABCDEF.

The string in the last example in Table 7.2 could look like:

AAAAAAAAABBBBBBBBBBCCCCCCCCCCDDDDDDDDDDEEE
EEEEEEEFFFFFFFFFF

but also could appear as:

ABCDEFABCDEFABCDEFABCDEFABCDEFABCDEFABCDEFAB
CDEFABCDEFABCDEF

or any order for that matter — as long as each letter appears 10 times in the total string of 60 letters. For each of these cases, we can calculate how many bits we need as a minimum to achieve a binary encoding of the string, using the method of the Shannon entropy outlined above.

The results can be seen in Table 7.2. The Shannon entropy follows the same trend we found above: the less predictable each successive letter in a string, the more bits are required to encode the entire string. For instance, the first case will require 60 x 0.61 = 36.6 bits, whereas the last case will need 60 x 2.58 = 154.8 bits. To encode six different letters in binary digits, we need a three-digit binary code (since 2^3 = 8 coding possibilities, since two digits would give you only four (2^2 = 4) coding possibilities). Thus, if we do nothing else, we'll need 60 x 3 = 180 bits to encode each of the 60-letter strings in Table 7.2.

But, you might argue, the Shannon entropy indicates substantially lower numbers of bits required to encode. What's going on? The answer is data compression. Now welcome back to the principle of "minimum bits required," as calculated with the Shannon entropy. Data compression is a smart way to encode the string, but with fewer digits than the three per character we would need otherwise. We will see in a moment that data compression is based on the frequency characteristics of the letters. The Shannon entropy shows the limit of the data compression; in other words, no matter how clever your data compression algorithm, you can never get a compression better than the Shannon entropy[121].

The last two columns in Table 7.2 illustrate this idea. The "Possible Compression" column shows the difference between the minimum number of bits required (the Shannon entropy) and the number of bits needed to encode without any data compression (3), then divides by 3 to get relative compression. For instance, in the first row this works out as $((3-0.61)/3)100 = 80\%$. Therefore, the Possible Compression column indicates the maximum amount of compression possible, or a best case, so to speak.

Table 7.2 Compression of a string of 60 letters

Occurrence of letter in string						Shannon Entropy (bits/letter)	Possible Compression (%)	Shannon Fano Compression (%)
A	B	C	D	E	F			
55	1	1	1	1	1	0.61	80	31
50	5	2	1	1	1	0.98	67	31
40	7	5	4	2	2	1.64	45	26
30	8	7	6	5	4	2.14	29	21
20	10	9	8	7	6	2.45	18	17
15	10	10	10	8	7	2.54	15	14
10	10	10	10	10	10	2.58	14	11
Total length of string is 60 characters								
For 6 characters we need 2**3, is three bits per character								

---

[121] This is true as long as we consider what is called "lossless" compression. These are compression techniques that will not lose any detail of the original message. In other words, it is always possible to restore the original exactly from decoding the compressed signal. This is, of course, very important in the case of written messages and databases. An example of a lossless compression technique is that of TIFF (Tag Image File Format) for images. With music and moving images it's possible to take advantage of human perception. In those cases much more compression is possible (for example in MPEG and MP3 compression, see below), but the price paid is some lost information. The original cannot be exactly restored. These techniques are called "loss compression".

7: The Use of the Concept of Entropy in Other Sciences

Table 7.3 Probabilities of the letters in the alphabet for the English language

Letter	Probability of Use in English Text	Code 1	Code 2	Code 3
(Space)	0.1859	00000	00	000
A	0.0642	00001	10100	0100
B	0.0127	00010	0111100	011111
C	0.0218	00011	01101100	11111
D	0.0317	00100	011100	01011
E	0.1031	00101	100	101
F	0.0208	00110	1101100	001100
G	0.0152	00111	0101100	011101
H	0.0467	01000	111100	1110
I	0.0575	01001	1100	1000
J	0.0008	01010	101010100	0111001110
K	0.0049	01011	0110100	01110010
L	0.0321	01110	01100	1001
M	0.0198	01101	010100	001101
N	0.0574	01110	01100	1001
O	0.0632	01111	01010100	0110
P	0.0152	10000	10101100	011110
Q	0.0008	10001	010110100	0111001101
R	0.0484	10010	101100	1101
S	0.0514	10011	11100	1100
T	0.0796	10100	0100	0010
U	0.0228	10101	1100100	11110
V	0.0083	10110	1110100	0111000
W	0.0175	10111	1010100	001110
X	0.0013	11000	01110100	0111001100
Y	0.0164	11001	011010100	00111
Z	0.0005	11010	01011100	0111001111

(Data from www.lucent.com, used with permission of Lucent Technologies)

The principle of data compression can be illustrated with an algorithm proposed by Shannon and Fano. The cases in Table 7.2 allow us to explain the algorithm pretty easily. For example, let's take the third case:

Character	A	B	C	D	E	F
Frequency:	40	7	5	4	2	2
Uncompressed coding	000	001	010	100	110	101
Shannon/Fano compr.[122]		10	01	010	100	110
						101

---

[122] I have to add here that there are more coding possibilities, because with three digits we have eight possibilities to code, but we need to code only six letters.

For the uncompressed coding we need: 40 x 3 + 7 x 3 + 5 x 3 + 4 x 3 + 2 x3 + 2 x 3 = 180 digits, which we saw above. The Shannon/Fano compression, however, needs only: 40 x 2 + 7 x 2 + 5 x 3 + 4 x 3 + 2 x 3 + 2 x 3, or 133 digits. Thus the compression is (180-133)/180x100 = 26% (which is the value that can be found in the table as well). In this way, we can calculate the actual compression achieved using the Shannon/Fano compression algorithm and compare it with the best compression possible, according to the Shannon entropy. What we see is that none of the actual compressions is better than the best possible one defined by the Shannon entropy.

In Table 7.3, we can see what the actual frequencies are for normal English text, as well as three lossless[123] coding techniques. The Shannon entropy calculated for these frequencies is 1.5 bits. The compression results for code 1 is 5 bits/letter, while code 2 needs 4.9 bits and code 3 requires 4.1, still far from the ideal case of 1.5 bits/letter. Indeed, other very advanced compression techniques (also made possible by today's faster computers) have been designed that can bring the achievable compression much closer to the ideal one. This can be achieved, for instance, by recognizing that a certain combination of letters (say, "th") is more likely to be followed by a vowel than a consonant. Other examples of modern compression techniques are MP3 for music and MPEG for moving images (but see my remarks in footnote 121). MP3 can reduce the bit stream for audio fragments from about 1400 kbit/s for an uncompressed CD quality to about 128 kbit/s (more than a factor of 10), while MPEG can lower the bit stream for video images from 128 Mbit/s by a factor of 10 to 100, depending on whether stored or live image data is transferred. The science of data compression is still developing, because of our ever-present hunger for more data transfer and ultimate desire for negligible transfer times.

Now that we understand a little better how the Shannon entropy was introduced and used in communication theory, I can guess your next question: What does Shannon entropy have to do with Boltzmann entropy? There are two answers, neither of them simple. The first is purely mathematical, and starts with the Boltzmann formula used in Chapter 3:

$$S = k \ln W \qquad (3.1)$$

---

[123] See footnote 121.

7: The Use of the Concept of Entropy in Other Sciences 149

with k being the Boltzmann constant and W the amount of microstates that form the macro state for which we want to calculate the entropy. Now we compare equation (3.1) to Shannon's formula:

$$S = -K\sum_{i}^{N} p_i \log_2 p_i \qquad (7.1)$$

Math nerds, and probably no one else, will recognize that equations (3.1) and (7.1) are identical — but believe me, they're right.[124] This equality led Shannon's friend John von Neumann[125] to advise Shannon to call his parameter "entropy." According to one anecdote, von Neumann told Shannon, *"Since nobody knows what entropy really is, it will give you an advantage in discussions."* Shannon followed his friend's advice and pointed out in his 1948 article that S "will be recognized as that of entropy as defined in certain formulations of statistical mechanics." He then referred explicitly to the work of Boltzmann, who derived an identical equation for the entropy, such that p_i is the probability that the system will have energy E_i.

The second answer involves a physical interpretation. From a physical point of view, Shannon's entropy has nothing to do with the entropy of Boltzmann. With the latter, we are dealing with distributions of atoms and molecules across different available energy levels in the macrosystem, and a continuous dynamic process that interchanges energy among atomic particles. Shannon's letters or bits do not behave like that at all[126]; they're not in constant motion, exchanging energy and colliding with one another. However, Boltzmann entropy can be seen as a measure of disorder in a system, or to put it differently, as a measure showing the probability of a certain potential state. Similarly, a sequence

---

[124] I apologize here that I am not more explicit, but the mathematical proof goes beyond the scope of this book. For the real diehards among you: key in the proof is the Stirling approximation that states that $\log (x_1 x_2 x_3 \ldots x_n) = x_n \log x_n - x_n$ if x_n, which is valid for large numbers.

[125] John von Neumann, born in 1903 in Budapest, Hungary, was a brilliant mathematician and since 1931 was connected to Princeton University. Although he achieved many contributions in diverse fields, he is probably best known as the father of computer science. He proposed automatic calculations while he was a consultant to the Los Alamos Scientific Laboratory from 1943 to 1955 where he suggested the implosion concept to construct nuclear arms.

[126] However, at the risk of confusing you even more, the discussion of Maxwell's demon later in this chapter shows that there may be more connection between Shannon entropy and Boltzmann entropy then we originally thought.

of letters involves a certain probability that a particular letter will appear next, and the less certainty we have of a precise letter appearing (i.e., more disorder), the higher the value of the Shannon entropy. So, there are perhaps even three ways to look at it: mathematically, there is a relationship between the two entropies; physically, they have nothing to do with each other; and disorder-wise, there is again some correlation between the two entropies. (For a further explanation, see the discussion on the entropy of a deck of cards in Appendix IV.)

Regardless of how the relationship between the two entropies is described, there is perhaps one more similarity: both had a tremendous impact on later developments. Boltzmann's work increased understanding of the physical meaning of entropy, which aided the development of relativity, quantum physics, and the foundation of the atomic world view (as we saw in Chapters 3 and 4). Shannon's work laid the groundwork for the new science of information and communication theory. That in turn enabled today's communications tools such as television, mobile phones, MP3 players, and satellite radio. These advances would not have been possible without concepts such as data compression, identified in Shannon's mathematical theory.

Maxwell's demon

Now that we've discussed Boltzmann's and Shannon's interpretations of entropy, let's look at one of the most famous thought experiments in history: Maxwell's demon. Why? That is because, as we will see below, in cracking the demon case a relation was laid between Boltzmann's entropy and information content, but in a rather different and unexpected way than above. Before we do that, let's spend a few words on thought experiments, which are very popular in science and philosophy.

Thought experiments are conducted in the mind rather than in reality. They allow us to come up with all kinds of situations that would be very difficult to realize in the real world. A famous example is Einstein's paradox of human twins: One twin is put in a rocket traveling at speeds close to light, while the other twin remains on earth. When the traveler returns home, one wonders whether they are still the same age. The answer is no, because the traveler has aged less quickly than his earthbound brother. Such is the value of a thought experiment,

7: The Use of the Concept of Entropy in Other Sciences 151

impossible to carry out but dramatically illustrative[127]. Thought experiments typically are invented to challenge new theories; sometimes it can take many years before scientists agree on the experiment's outcome.

Another good example of a thought experiment is Maxwell's demon. Originally proposed by James Maxwell in 1867 in a letter to his friend, the mathematician Peter Tait, it took almost 150 years before the discussion on this experiment concluded. Figure 7.1 shows the demon at work.

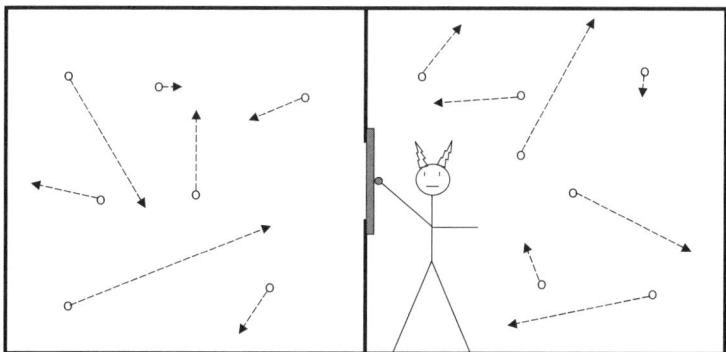

Figure 7.1 A sketch of the thought experiment created by James Clark Maxwell, also known as Maxwell's demon. The demon (or creature, as Maxwell called it originally), is able to open a frictionless door and determine the velocity of the individual gas molecules. The size of the arrow represents the velocity of each molecule. See text for more details.

Let's explain: First, imagine two chambers connected by a small hole in a common wall. The hole is covered with a frictionless door that is operated by a "very observant and neat-fingered being," as Maxwell described it. Both chambers are filled with an equal number of gas molecules at the same temperature. We have seen earlier (in Chapter 3) that at a given temperature, we can calculate the average speed of the gas molecules. However, we also know that there exists a wide distribution of speeds of gas molecules in the two chambers, and that the average velocity (speed in one direction) of the individual gas molecules is zero.

---

[127] Indeed, later experiments, in which atomic clocks in jets took the place of the hypothetical twins, proved the thought experiment to be right.

Now, imagine that the little creature is quite talented and can figure out the speed of each gas molecule that is approaching the hole in the wall. Every time a gas molecule with higher-than-average speed approaches from the right, the creature quickly opens the door. And every time a slow gas molecule approaches the hole from the left, the door is opened and quickly closed after the slow molecule goes through. Eventually, the left chamber will contain gas molecules with an average speed higher than the initial average, and the right chamber will contain gas molecules with a slower average speed. Thus, the temperature of the right chamber has dropped and that of the left chamber has risen, without any work having occurred. Even stranger, the creature is apparently able to move heat from the colder (right) chamber to the warmer (left) chamber. However, both observations violate the Second Law, and that's why the keen-witted being soon was named Maxwell's demon. One reason Maxwell conjured his demon was that he was unhappy with the consequences of the Second Law for the world, i.e., Heat Death. Although he was convinced that the Second Law was correct, he was looking for ways to outmaneuver it.

Several other versions of the demon were invented later, but all had the common flaw of violating the Second Law. In a simplified version, the demon no longer has to identify the speed and direction of the gas molecules, but need only open the door when a molecule comes from the right. In this case, the right chamber will hold fewer molecules over time, while the left chamber fills up. Thus, the demon acts as a one-way valve. Of course, this will result in a pressure difference between the two chambers and that pressure difference then can be used to drive pneumatic equipment. This example essentially makes the demon a perpetual engine, which we discussed in Chapter 2. There we learned that as long as a perpetual engine stays on the drawing board, it can be extremely difficult to show where its flaw resides. Similarly, Maxwell's demon might very well be history's most difficult thought experiment to disprove. Since its conception in 1867, many brilliant scientists (Brillouin [1950], Szilard [1927][128], and Smoluchowski [1914], to name a few) have come up with proofs (or "exorcisms," as some publications have called them) why the demon would not work. One proof asserted

---

[128] Leo Szilard, a Hungarian, in 1927 at the University of Berlin devoted his thesis "On the increase of entropy in a thermodynamical system by action of intelligent beings" to Maxwell's demon. It was here that the relation between entropy and information was first established.

7: The Use of the Concept of Entropy in Other Sciences 153

that the demon would need a flashlight or other light source to determine the speed of the approaching molecules. Some photons would be scattered by the molecules and would be lost, thus leading to a change in entropy. Another proof held that because of Brownian movement, the door would heat up (the hole and the door being very small), from gas molecules continuously colliding against it, and it would cease to work as a one-way valve. Although the demon was cast out many times by these proofs, it was always able to scramble back because others found flaws in the proofs. The discussions continue to this day, 140 years later.

An interesting and unexpected insight into the discussion was offered by IBM scientists searching since 1960 for the fundamental limits of the process of computing [Bennett and Landauer, 1985]. The inspiration for this work came from thermodynamic analysis of the efficiency limit of a steam engine, and from Shannon, who gave engineers a quantitative expression for the information content in a signal. If we add two numbers, such as 1 and 3, the result is 4. However, once we have obtained the 4, we no longer know which numbers added, since 4 can be the sum of many numbers. So from that perspective, information is lost. Numbers in computers are often stored in memories. Erasing a bit in a memory means that information is lost, since afterward we no longer know the values of the previous bits. For instance, a bit can have the value of 0 or 1. After erasure, we no longer know its value, and we have increased the information entropy by $\Delta S = k ln 2$. At the physical system level, erasing is done by moving electrons from the different memory cells (a memory cell is typically a combination of a capacitor that holds the charge, and one or more transistors that act as on and off switches). Moving the charges around in the memory will be accompanied by a certain amount of energy dissipation, ΔE. However, it appears now that such energy dissipation can never be made smaller than $T\Delta S$, which establishes a fundamental efficiency barrier. But let's not forget that present-day computers are about eight orders of magnitude above this theoretical limit of energy dissipation, enough room for improvement I would say. And indeed, a lot of research is done these days to improve on this phenomenon driven by the need for ultra low power microelectronics in order to extend battery life time [see for a nice overview on the thermodynamic limits of computing Frank, 2002].

Let's revisit our demon. Some scientists suggested that in order for the demon to function, it needed to memorize some aspects of the approaching molecules, such as direction and speed, and could then decide whether to open the door. If we want to have a cyclic (and thus workable) process, the demon has to purge its memory from time to time in order to capture the direction and speed of new molecules. This erasure, as we have seen above, doesn't come free and in fact increases the entropy of the closed system that the demon is part of. Thus, the entropy of the system decreases with each molecule that passes from one chamber to the other. However, because the demon has to erase its memory, the entropy will increase by an equal amount, and so will result in zero entropy change. This now is seen as yet another proof that the demon proposed by Maxwell will not be able to violate the Second Law.

Use of the concept of entropy in other nonscientific fields

Entropy in the discussion of Christianity

The relationship between the Christian religion and science has often been difficult. We all know of the examples of Galileo Galilei, and the ongoing battle between Christian creationists and proponents of Darwin's theory of evolution. Another, more obscure battlefield is that of the Christian religion and the Second Law.

In Christian faith, the two Laws of Thermodynamics are considered the most fundamental laws in our world. Bible experts believe that both laws are mentioned in the different books of the Bible. For instance, the law of conservation of energy is believed to be implicit in Genesis 2:13: *"Thus heaven and earth were finished, and all the host of them. And the seventh day God ended his work."* The Second Law is often identified with death and decay, which are thought to be related to the sins of Adam and Eve: *"Cursed is the earth and thy sake"* (Genesis 3:17). Thus, the First Law does not cause any problems with respect to the interpretation of the Bible. It is the Second Law that is (still) the subject of a lively debate between creationists and scientists, but the quality of that debate fluctuates. There are two reasons for this: first, the concept of entropy is often very poorly understood and incorrectly cited; and second, entropy is identified with the bad things in life.

In the creationist's view, God is creator of life and life is progress toward more order. Here is where the problem comes in, since the non-scientists interpret the Second Law as requiring a continuous

7: The Use of the Concept of Entropy in Other Sciences 155

increase in disorder. In their view, there must be a power able to counteract the Second Law, which is cited as a proof of God's intervention. A similar problem arises with evolution theory and the Bible's interpretation. Evolution can be defined as a continuous process in which organisms change their genome through spontaneous mutations in the DNA and undergo subsequent natural selection, and as such increases the amount of order. Therefore, evolution is seen as being in conflict with the Second Law. On the Internet, you can easily find lots of websites where these topics are discussed or explained. Three types of people participate in these discussions: faithful people who have no scientific (or thermodynamic) background, and who therefore often misrepresent the Second Law; faithful people with a solid scientific background; and non-believers who are also scientific. Personally, I believe that the second category of people often provides the most balanced view on the matter. However, the discussion takes place in a somewhat fragmented fashion. I believe that it is therefore advantageous to point out a few lines of thought as they appear in the discussions.

The most common error is that the system of study is not properly defined, which constitutes (in ecclesiastical language) a mortal sin, since it will lead to absolutely wrong conclusions. I restate here that the Second Law says that the entropy of an *isolated* system will always increase for spontaneous (irreversible) processes. The key here is the word *isolated*. Isolated means that the system will not exchange energy or material with its environment. However, organisms are not isolated systems; even the earth (as we have seen previously) is not an isolated system. Both organisms and the earth have a continuous influx of energy. As long as this is the case, it is very possible that the entropy of such a system goes down.

But even in an isolated system, it is very likely that in some part of the system, the entropy decreases locally while the entropy of the total system increases. Let me illustrate that with a simple example in Figure 7.2 (and another example can be found in Appendix V).

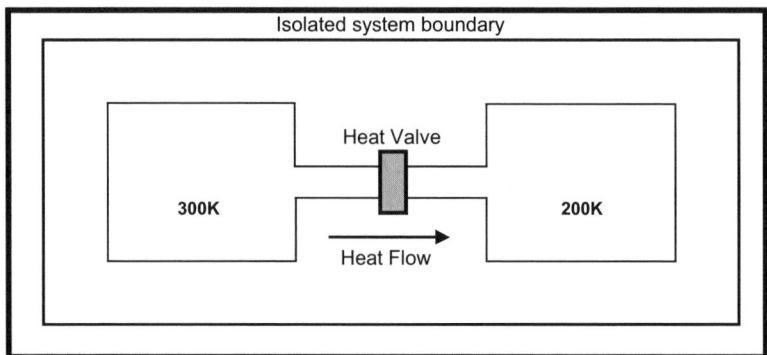

Figure 7.2 Two heat reservoirs in an isolated system are connected through a heat valve. The heat valve can be opened and closed. When the valve is open, heat will flow from the warm to the cold reservoir.

Imagine that we have a hot and a cold block of material in an isolated enclosure, as in Figure 7.2. Again, *isolated* means that no energy or material can be exchanged with the surrounding environment. Therefore, to understand the behavior of this system, we only need look inside the enclosure. We can connect the two blocks with a heat-conducting bridge equipped with a heat valve. The heat valve has two states: when it's on, it conducts heat, and when it's closed, it completely isolates the two blocks (for simplicity's sake, we assume that no other heat conducting mechanisms are possible). If we open the valve to let some heat through, we know that the heat will flow from the hot block to the colder block. Let's assume that a small amount, ΔQ, will flow such that the two temperatures do not change, and that we then close the heat valve. What will this mean for the entropy changes in the two blocks, and for the entire system? This is not a difficult problem to solve, because we remember from Chapter 2 that $\Delta S = \Delta Q/T$, i.e., the change in entropy equals the change in heat divided by the temperature. So let's utilize this simple formula as follows:

Entropy change for hot block: $\Delta S_{300} = -\Delta Q/300$
Entropy change for warm block: $\Delta S_{200} = \Delta Q/200$

7: The Use of the Concept of Entropy in Other Sciences

Entropy change for total system:

$$\Delta S_{Total} = \Delta S_{300} + \Delta S_{200} =$$
$$= \Delta Q(1/200 - 1/300) =$$
$$= \Delta Q(0.005 - 0.0033) =$$
$$= \Delta Q\ (0.0017)$$

First of all, we see that ΔS_{Total} is a positive number! Thus the Second Law is not violated but, perhaps most importantly, it is true only for the *isolated* system. For non-isolated systems (such as living organisms) that exchange both energy and materials (food) with the environment,[129] there is no conflict whatsoever with thermodynamic law, and the entropy can certainly *decrease* in such systems! But let's have a further look at the system in Figure 7.2. The entropy in part of the system *did* decrease (ΔS_{300} is a negative number, meaning that entropy is lost) but this loss is compensated by the entropy gain of the cold block (ΔS_{200} is a positive number). So, thermodynamics is not invalidated by a local decrease in entropy. If you properly define the system boundary under study in terms of isolated, closed or open systems, the true picture will emerge.

Another problem area is the perceived determinism of the Second Law. The Second Law is seen by some as something that would take away the possibility of choice for humanity, so that the entire universe and life itself would be predetermined. Again, the same error is being made. If we consider our planet as the system under study, then we know that because of the influx of energy from the sun, there is plenty of room for a bi-directional trend for the entropy. Of course (as we have seen above), when the entropy decreases in one location, there must be a compensatory increase elsewhere, but the vast scope of the universe makes that a minor concern to we humans here on planet earth.

---

[129] I need to point out here that the Nobel Prize winner, Ilya Prigogine, calls systems that do exchange energy with their environment dissipative systems. He shows that for such systems, which are not necessarily living organisms, a trend toward more order can occur. This order is qualitatively visible in, for instance, structures like the genome or the growth of organs from cells. However, it is fair to say that Prigogine's ideas are not shared by all scientists, and have openly been debated because his theories lack predictable results.

The concept of Entropy and art

Art can be described essentially as a communication between an artist and a spectator. We have seen how Boltzmann made a connection at the microscopic level between entropy and the degree of order of a given system, and how Shannon linked his version of entropy with the amount of information in a message and, inherently, the amount of redundancy in it. Also, we have explored the relationship between entropy and the arrow of time. These connections have inspired some artists to incorporate the concept of entropy in their work. We'll look at a few examples of how the entropy concept has been incorporated in works of art, but first I'd like to show you a few interesting observations by Rudolf Arnheim in his book, *Entropy and Art* [Arnheim, 1971].

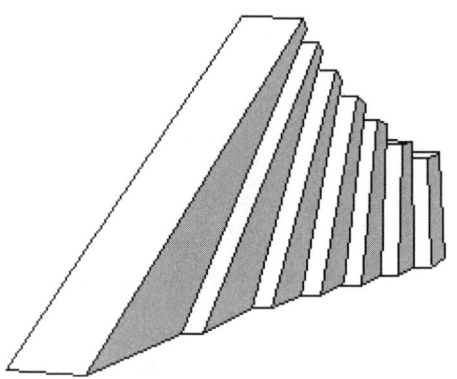

Figure 7.3 Sketch of "Leaning Strata" made by Robert Smithson in 1968 to create an impression of space.

He points out that while the First Law was a token of a timeless permanence (energy cannot be created or destroyed) and was interpreted as an evidence of "God's presence and action," the Second Law was connected to the somber *fin de siecle* mood of degeneration and the approaching end of the world (Heat Death). For instance, the indistinct perception of the Impressionists was to some observers a sign of degeneration and cultural degradation. Then there is the amount of information within a work of art, such as a painting. While information

7: The Use of the Concept of Entropy in Other Sciences

theory tries to find ways to reduce redundancy as much as possible (via data compression, as previously discussed), an artist can use redundancy as an essential part of the painting. A lot of redundancy in a signal lowers its entropy. A painter, however, can decrease the Shannon entropy of a painting, and in that way counteract the universal trend of continuous entropy increase! Influencing the laws of nature with a painting—isn't that something?

However, only a very few artists have explicitly mentioned that they made art while under the influence of the entropy concept. Of course, the connection between entropy and order/disorder made the interpretation of entropy by artists much easier than did the strictly classical thermodynamic interpretation of Clausius, with his emphasis on heat flows.

One artist who very clearly explained how he was influenced by entropy was Robert Smithson. Smithson was active as a versatile artist in New York from 1959 until his accidental death in 1973. Smithson was very much interested in certain sciences such as physics (notably crystallography) and geology, and tried to make connections between his art and concepts from those sciences. One topic that particularly drew his attention was entropy, and he seemed to understand the concept rather well. He argued that entropy through the Second Law was valid for all materials and living things on earth, and thus also for art. He wrote several articles about his interpretation of entropy and its connection to his art. This connection manifests itself along three lines. First, entropy can be seen as a general law that causes deterioration and decay. This led Smithson to use materials in his art that were already used and considered worthless. In addition, when he made outdoor sculptures, he took abandoned sites (for instance, old salt mines) and tried to "re-use" the land by converting the abandoned site into a new use. Second, since entropy has a connection to time, it connects also to past and future. Smithson believed that art must be dynamic, and that there must be a relationship between art and time. For instance, he and other artists made pieces that had only a limited lifetime and would change appearance during their existence. Third, there was a connection between entropy and communication and information theory through the work of Shannon. Indeed, parallelisms have been made between Shannon's transmission and reception of a signal, and art's function as another kind of signal [Bijvoet 1997]. If art is considered a signal, the element of time returns, since a signal always elapses in time.

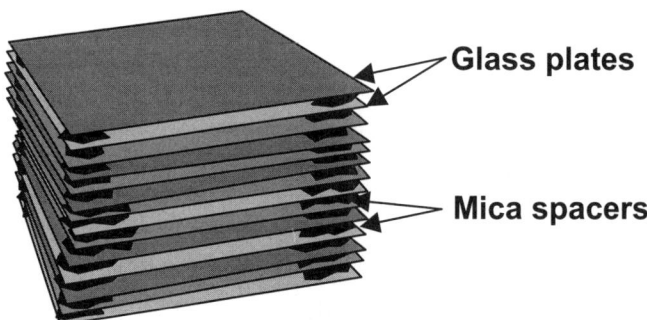

Figure 7.4 Sketch of the sculpture "mica and glass" made in 1969 by Robert Smithson. In this work the artist united amorphous (glass) materials with crystalline (mica) materials.

Another angle that Smithson drew from his entropy considerations was that of order and disorder. He worked a lot with natural materials, such as salt and glass. He was intrigued by the fact that glass could be given very regular shapes (for instance, rectangular plates, which he used quite often in his series of Strata sculptures, as in Figure 7.3). But at the microscopic level, the glass molecules are unstructured, since glass is an amorphous material. Salt, however, is at first sight a disordered material. But at the atomic level, we know how regular its molecular structure is, because salt is a crystal. In Figure 7.4 we see another example of how Smithson arranged (amorphous) glass plates separated by pieces of (crystalline) mica[130].

---

[130]There are many mica materials but they all contain silicate (SiO_2) tetrahedrons and elements such as iron, magnesium, potassium, lithium, and aluminum.

Epilogue

It started all with Sadi Carnot. Driven by the fact that France was behind in the industrial application of steam engines and early perceptions that these machines seemed to have unlimited efficiency, he did a careful analysis of heat flows at the engine's different temperatures. That was the beginning of thermodynamics, the science of heat and work. This science led to two somewhat paradoxical laws: the First Law, or conservation of energy; and the Second Law, or entropy. Whereas the First Law says energy cannot be created or destroyed and, in a sense, gives immortality to energy, the Second Law predicts a finite life for our universe, our planet and, unfortunately, for us.

We have seen how, based on purely empirical observations, a framework can be built leading to the relatively simple and elegant theory called classical thermodynamics. (The simplest of these observations is that heat flows from hot to cold regions.) Although classical thermodynamics delivered a lot of successes, further refinements changed it from a theory based only on principles to a constructive theory founded on the statistical mechanical work of Boltzmann around 1900. Boltzmann's work allowed the scientific world a much closer look at the fundamental reasons for irreversibility, and this has fueled many discussions (for instance, on the direction of time) to the present day.

The development of thermodynamics occurred over what is probably the most interesting period in the evolution of science, when around 1900 the atomistic view of matter started to gain ground. While Carnot and his successors, including Mayer and Joule, only worked with macroscopic parameters such as pressure and temperature, Boltzmann believed very much in the reality of atoms and gave thermodynamics its statistical interpretation and extension.

Much of the energy wasted in modern industrial processes and economies is the price we pay for speed (of course, we now understand the connection between the degree of irreversibility and this speed problem). This introduces an interesting discussion about the productivity of processes. Intuitively, we define productivity as the amount of energy used per produced unit, or perhaps the speed with which we can produce an amount of output. The latter, of course, is connected to the amount of units we can produce per hour. However, if we considered this situation from a thermodynamic point of view, we would define thermodynamic efficiency in terms of the amount of entropy produced per unit of output. Every unit of entropy that we produce today will be one unit less for our children tomorrow, since the total amount of entropy that can be produced is limited. Indeed, this gives us a different perspective on the problem of world resource conservation and utilization.

Much to our surprise, we have seen that the Second Law is related to a variety of seemingly unconnected issues, such as the flow of heat, the direction of time, the characteristics of life, the secrets of the universe, the wear and tear on your painted walls, and the microscopic processes running in computer memories. In all these cases, we are dealing with a direction, and the Second Law only tells you which way to go, but not when or how something will happen. This aspect of the Second Law we can appreciate, because it keeps life interesting and full of surprises!

In summary, thermodynamics is a convincing illustration of the power of the human mind when applied to a very fundamental field, namely the understanding of energy transfers. As Einstein has said, it is safe to assume that this theory will live on forever, and will continue to provide our society with insights and guidance in energy conservation and many other areas.

Appendix I. Two More Laws of Thermodynamics?

Sometimes, you run into scientific works that discuss two more laws of thermodynamics: the Zeroth Law and the Third Law. There has been intense debate in professional literature whether these two laws are not already implicit in the First and Second Laws. But we won't concern ourselves with that discussion, so that we can learn more about the rationale for these two additional laws.

The Zeroth Law considers three bodies A, B, and C. Bodies A and B are in contact so that heat can flow between them. After a certain amount of time, we assume that A and B have reached the same temperature and so are in *thermal equilibrium*. The same thing occurs when A and C come into contact and remain that way for a while. We repeat this till the temperature of A does not change anymore when we bring it contact with either B or C. The Zeroth Law states that when A and B and A and C are in thermal equilibrium, then B and C also are in thermal equilibrium and thus all three bodies will have the same temperature.

The Third Law deals with the value of entropy at absolute zero. We have seen in Chapter 2 that $\Delta S = \Delta Q/T$. When the temperature goes to zero, then the entropy change with a certain amount of exchanged heat becomes infinitely large. This is not what has been observed experimentally, and the opposite seems to be true: at temperatures close to absolute zero, it appears that the specific heat capacity[131] of materials approaches zero as well. Therefore, ΔQ will approach zero, which will make the entropy change zero (W. Nernst first formulated this law).

From a molecular point of view, one also can understand that the entropy is zero at zero absolute degrees. In the case of a crystalline solid, all atoms will be at the lowest possible energy, trapped in the lattice with no uncertainty left. That produces only one microstate to realize the macrostate at zero degrees, and thus the entropy, according to Boltzmann's expression, becomes zero.

---

[131] The specific heat capacity is the amount of heat needed to increase the temperature of one gram of material with one degree. Its unit is accordingly J/g K.

Appendix II. Another Way of Looking at Entropy

The discussion in Chapters 2 and 3 hopefully has given us good insight into how scientists discovered the concept of entropy, and how this concept can be used to explore the world of energy transformations and heat transfers. However, I still believe that some of us will be left thinking that there must be something more to it. For that reason, I would like to offer a slightly different way to consider entropy, and how it determines the directions that processes in the universe must take.

We have seen that pressure, volume, and temperature are important thermodynamic parameters which characterize the state of a given system. However, there is an important feature common to those parameters. Imagine that we have two containers, each of a volume V and connected by a hose and a two-way valve. Each container contains a gas at a certain pressure and temperature. What happens when we open the valve? Not much, you might say, but there is a fundamental difference between the pressure and temperature on one hand and the volume on the other. After opening the valve, the volume of the system becomes 2V, but the pressure and temperature do NOT increase, and instead extend their former values to the entire new volume. Scientists, therefore, call pressure and temperature *intensive* parameters, while volume belongs to the class of *extensive* parameters. It was the famous scientist Josiah Willard Gibbs who, working at Yale University on heat problems while Clausius was doing the same in Germany, defined a general expression for the calculation of energy changes. This formula, called the Gibbs fundamental equation, states that the change in energy can always be written as the product of an intensive and an extensive parameter:

$$\Delta U = \sum_i X_i \Delta Y_i \quad (A2.1)$$

In formula (A2.1), X_i is an intensive parameter whereas Y_i is an extensive parameter. If we take a given volume of gas, then we feel intuitively that pressure and temperature characterize the total state of that volume of gas, and must play an important role in the description of energy changes that that volume of gas can undergo. We have seen earlier, in Chapter 2, that $\Delta U = \Delta Q + \Delta W$, or that both heat and work can change the energy content of a system. For the work component, we feel very comfortable in replacing ΔW with the term $P\Delta V$. However, for the change in heat we must have an intensive parameter, and since we already used P for the work amount, all that's left is the temperature, T. Thus, ΔQ can be equal to T times something, and that something must be an extensive property. This property is called the entropy, S, and so we can write $\Delta Q = T\Delta S$. The idea, then, is that at zero pressure there will be no work, and at zero temperature no heat flow will be possible. From simple calculations it is indeed easy to show that after the valve is opened, the entropy of the system is the sum of the two entropies that each volume of gas had before the valve was opened (plus a bit more, because the gas molecules can achieve more possible states). See also Appendix V for further discussion.

Appendix III. How Does the Gas Heat Up in the Air Pump?

We all know that if we use an air pump to inflate a tire or cushion, the temperature of the pump will increase at the outlet. The reason is that the temperature of the air increases when it is compressed by the pump. But why is that? Well, from a macroscopic thermodynamic point of view that's pretty clear, as we have seen in Chapter 2. With each downward stroke of the plunger, we deliver an amount of work[132], and that amount is partly transformed into heat. Why? Well, assume that the downward stroke is rather quickly done. That means that the compression is adiabatic, no heat can be exchanged with the environment. Thus $\Delta Q = 0$ and $\Delta E = -P\Delta V$. Thus we have an increase in energy because ΔV is negative (compression) and we have seen in Chapter 2 that the energy of a gas depends on its temperature. Therefore, dissipated heat in the gas because of the work done is then expressed as a rise in temperature. No surprises so far — but what's really happening at the atomic level? We know that temperature basically is produced by fast-moving atoms. Or to put it more precisely, the average velocity of the atoms and molecules in a gas is directly related to the temperature of that gas.

But what causes the velocity of the gas molecules to increase when we compress the gas? After some thought, we may conclude that it's because the piston in its compression stroke has a certain velocity (let's say 1 meter per second). During the compression stroke, the piston will hit many gas molecules and increase their velocity (like hitting a tennis ball with a racket). Well, you're on the right track, but the situation is actually a bit more complex.

For the right answer, we first must consider the average velocity of nitrogen (the main constituent of air) at room temperature: that's about 450 m/s. So, the plunger with its speed of 1 m/s does not seem to have much of an impact, since it will only increase molecular speeds from 450

---

[132] Remember, the work done when a gas is a little bit compressed is the change in volume times the pressure or $P\Delta V$. Of course when the change in volume is big, as we have here, the pressure changes as well during the compression and a calculation of the work done is a bit more cumbersome but stays in essence the same (for instance we can split up the total compression in many little steps and keep the pressure in each of the little steps constant. This process is called in mathematical terms integration).

to 451 m/s. Obviously, there's more to it. So let's take a container with 10 liters of gas at room temperature (293K) and one atmosphere of pressure. Under those conditions, it can be calculated that there are about 3.10^{23} collisions per second per square centimeter (cm^2) of nitrogen molecules against the walls of the container. In fact, these collisions cause the pressure of the enclosed gas! The container is provided with a piston that has a surface area of 100 cm^2, and thus will have 3.10^{25} collisions per second. Assuming that we compress the volume with one liter, our piston will travel for 10 cm in 0.1 sec (remember speed was 1 m/s), experiencing a total of 3.10^{24} collisions. Since the container holds about 3.10^{23} molecules of nitrogen, this means, therefore, during the compression stroke each molecule will hit the piston an average of 10 times ($3.10^{24}/3.10^{23}$). These 10 collisions will increase the average velocity of the molecules to 460 m/s. Now, the temperature of a gas molecule depends on the square of its velocity. Therefore, if an average velocity at 293K is 450 m/s, then at 460 m/s the temperature must be:

$$\frac{(460)^2}{(450)^2} x 293K = 304K$$

Thus, the temperature of the gas in the cylinder has been increased by about 11 degrees. Although our calculation is only approximate (we neglected, for instance, to include the increase in pressure from compression, which also increases temperature), it still helps explain the mechanism of gas heating by compression at an atomic level.

The fact that gas heats up during compression is of eminent importance and of great practical use. An example is the diesel motor, invented by Rudolf Diesel[133] in 1892. Diesel learned about the low efficiency of steam engines and tried to invent an engine that would improve the efficiency. He designed an engine that ran on petroleum and would not need an electrical ignition (as was the case with the Otto engine that ran on gasoline). In 1895, a working diesel engine was produced. In today's diesel engine, the high temperature needed to ignite

---

[133] Rudolf Diesel was born in 1858 in Paris. In 1892 he obtained a patent in Germany (DE67207) on his engine followed by several others in the years after. He died in 1913 under mysterious circumstances during a boat trip from Antwerp to London [Dulken 2000].

the fuel is obtained from a rapid compression[134] in which the compression ratio can go up as high as 20:1 or more. At a compression ratio of 20:1, the gas heats up to about 700°C and then a bit of fuel is injected and burned instantaneously. Fuel efficiencies can be as high as 60%.

---

[134] Because the compression is done so quickly (about 0.1 second), it can be considered as an adiabatic compression. This means that nearly all the work done during the compression is converted into heat allowing a rapid increase in temperature.

Appendix IV. Will Reshuffling a Deck of Cards Change the Entropy?

Imagine you have a neatly ordered deck of cards. Now you throw all the cards into the air and they fall back to the table in random order. Will the entropy have changed for the deck of cards? Absolutely not! The change in entropy is exactly zero. But wait a minute, you argue, the chaos of the deck has increased, and so should the entropy. To better understand why this isn't so, we need to backtrack a bit to Chapter 3.

Although not explicitly mentioned in Chapter 3, Boltzmann used statistical methods in his theoretical derivation of the entropy equation. Because of the enormous amount of atoms he had to deal with, it was impossible for him to consider each individual atom, so he had to rely on averages and determine the averages via statistics. Crucial to his approach was the distribution of the individual atoms over the available energy levels, and the dynamics of that distribution. The dynamics occur because the atoms have a momentum (speed times mass) which is determined by the temperature of the macrostate (the square speed of an atom is proportional to the temperature). This is, in fact, the reason that gas molecules will expand when given more space. Because the atoms collide continuously with each other, they exchange energy constantly, assuring that the available energy levels are occupied according to the so-called Maxwell-Boltzmann distribution. Only for such a system is Boltzmann's expression valid. In this example, it's clear that the individual cards in the deck have no interaction, no collisions, and no energy exchange whatsoever. Even if you allow the cards to take up more space by tossing them onto a larger table, nothing will happen, because there is no dynamic interaction going on.

Of course, when we use the entropy definition of Shannon, as discussed in Chapter 6, we do have a change in entropy—but this change is unrelated to thermodynamic entropy (or heat exchanges). More detail can be found in a discussion by Lambert [Lambert, 1999].

Appendix V. How Much Does the Entropy Change in a Case of Gas Expansion and Gas Mixing?

Gas molecules or atoms are in continuous motion, driven by their thermal energy. From the kinetic gas theory, we know that the thermal energy of an atom or a molecule is equal to $3/2kT$ (where k is Boltzmann's constant and T is the absolute temperature). The kinetic energy is $1/2mv^2$ and thus: $1/2mv^2 = 3/2kT$; where m is the mass of the molecule and v is the speed of the molecule. Whenever a concentration gradient arises, the gas molecules will try to reduce this gradient until it disappears and the concentration is the same everywhere. They can do this because of their thermal velocity. To get a feeling: at room temperature, a nitrogen molecule (the major constituent of air) has a velocity of about 450 m/s. The process that describes mass transport driven by a concentration gradient is called diffusion. Diffusion plays a role in almost every aspect of everyday life. For instance, the dissolution of the sugar cube in Chapter 2 is an example of diffusion, as is the vaporization of a water droplet in air. As we have seen in Chapter 5, the mining of certain ores and the subsequent liberation of the metals from the ores can be considered as the artificial opposite of a spontaneous diffusion process.

Using the diffusion process of gases, we can very easily show a few fundamental aspects of thermodynamics. For instance, the entropy in an isolated system always increases during a spontaneous process, and we can quite easily show the essence of what makes a process reversible or irreversible. Figure A5.1 shows two containers of volume V_1 and V_2, connected by a pipe equipped with a valve. The containers are completely isolated from the environment, with no energy or material exchange possible, and therefore constitute an isolated system. In the first case (top sketch) the left container is filled with n_1 molecules of gas and the valve is closed; the right container is completely empty. We open the valve slowly and the molecules quickly fill up the right chamber so that the pressure in both chambers equalizes. We have now a situation where n_1 molecules occupy the total available volume $V_1 + V_2$. Because

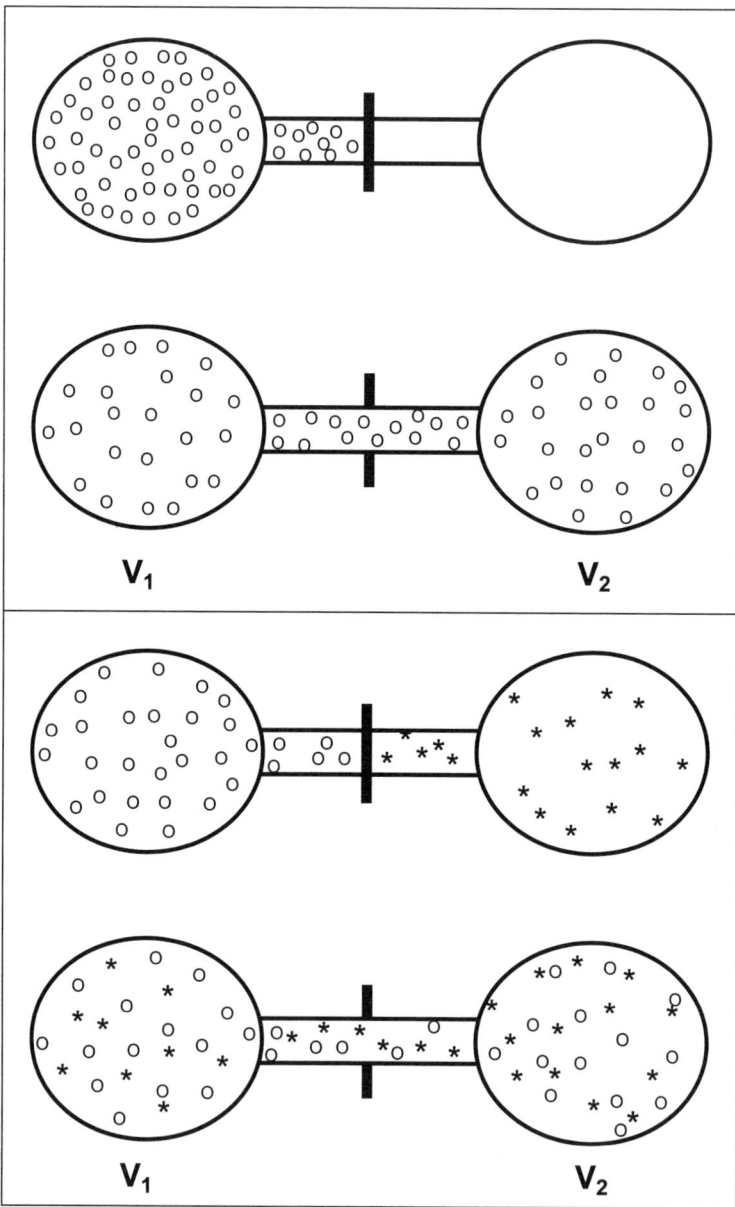

Figure A5.1. Two different cases of diffusion. Top figure shows the expansion of a gas, and the bottom figure shows the mixing of two gases.

the system was completely isolated from the environment, the energy has not changed. The question is, will there have been a change in the entropy? It is not difficult to answer the question. We begin with the fundamental differential equation[135]:

$$dU = TdS - PdV \qquad (A5.1)$$

where d denotes a very small change in U, S, or V. Because the system is isolated, we find that $dU = 0$ and therefore we deduce that $TdS = PdV$. In the case of an ideal gas, we can use the ideal gas law: $PV = n_1kT$ and thus equation (A5.1) can now be written as:

$$TdS = \frac{n_1 kT}{V} dV \qquad (A5.2)$$

and therefore:

$$dS = \frac{n_1 k}{V} dV \qquad (A5.3)$$

Since we expand the volume from V_1 to V_2, we must integrate[136]:

$$S_2 - S_1 = \int_{V_1}^{V_1+V_2} \frac{n_1 k}{V} dV = n_1 k \ln\left(\frac{V_1 + V_2}{V_1}\right) \qquad (A5.4)$$

This is an important result that shows a few things:
First, we see that $S_2 - S_1$ is *larger* than zero because $V_1 + V_2 > V_1$. Thus, since we're looking at a spontaneous process in an isolated system, this is an elegant illustration of the Second Law that the entropy will increase. In the simple case that the two containers have the same volume

---

[135] This equation is generally valid regardless of whether the process is reversible or irreversible. This may be surprising, since we have said earlier that only for reversible processes $\Delta Q = T\Delta S$ and that for irreversible processes $T\Delta S$ is greater than ΔQ_{rev}. Equation A5.1 is generally valid, however, because dU is a state function and independent of the route from the initial condition to the final condition.

[136] We must integrate because the difference between the start and end situations is large and V is therefore not constant but goes from V_1 to V_2.

($V_1 = V_2$), equation (A5.4) will reduce to $S_2 - S_1 = n_1 k \ln 2 \approx 0.7\, n_1\, k$.

Second, we are dealing here with an irreversible process. Why? Well, we don't expect that the expansion of the gas from the left chamber into the right chamber will reverse itself spontaneously. Or to put it more bluntly, it's unlikely that all the gas will retract into the left chamber while leaving a vacuum in the right chamber. This would, of course, violate the Second Law, because in such a process the entropy would decrease. We can even derive from this the general rule that spontaneous processes that show an *increase in entropy, are by definition irreversible.* Only spontaneous processes that have no entropy increase can be reversible.

Third, we observe that, despite having a completely isolated system with nothing (energy or material) added or subtracted, there is an increase in entropy! Thus entropy can be created within an isolated system (out of nothing, if you will).

Now let's have a look at the bottom situation in Figure A5.1. The initial condition is that we have n_1 molecules of gas 1 in the left chamber and another n_2 molecules of gas 2 in the right chamber. When we open the valve, we expect the two gases to mix until they are equally concentrated in the two chambers. Will this change the entropy of the system? To find out, we calculate by taking advantage of the *extensive* nature of the entropy, or to put it differently, we figure the total entropy by simply adding the entropies of gas 1 and gas 2. Thus, we only need to calculate the entropy difference after opening the valve between gas 1 and gas 2 and we're done! Here we go: the entropy change for gas 1 while expanding from V_1 to V_1+V_2 is actually given by equation (A5.4):

$$\Delta S_{Gas1} = n_1 k \ln\left(\frac{V_1 + V_2}{V_1}\right) \tag{A5.5}$$

Likewise, we can write expansion of gas 2 from volume V_2 to V_1+V_2 as follows:

$$\Delta S_{Gas2} = n_2 k \ln\left(\frac{V_1 + V_2}{V_2}\right) \tag{A5.6}$$

Therefore, we arrive at the total entropy change of the isolated system by:

$$\Delta S = \Delta S_1 + \Delta S_2 = n_1 k \ln\left(\frac{V_1 + V_2}{V_1}\right) + n_2 k \ln\left(\frac{V_1 + V_2}{V_2}\right) \quad \text{(A5.7)}$$

In the case where we have the same amounts of gas 1 and gas 2, thus $n_1 = n_2 = n$, and the two containers have the same volume, thus $V_1 = V_2$, we arrive at: $\Delta S = 2\ n\ k\ ln2 \approx 1.4\ n\ k$. We note that this is again a positive number, so we see an entropy increase and, therefore the Second Law is confirmed for this situation as well. In this case, there will be no pressure drop upon opening the valve (which is definitely the case in the expansion described above), and the entropy increase is solely due to the mixing of the two gases.

Appendix VI. Thermodynamic Timeline

1944	"What's Life?" published
1911	Quantization of black body radiation by Planck
1911	First Solvay Conference
1906	Formulation Third Law of Thermodynamics by Nernst
1905	Photoelectric effect explained by Einstein
1877	Statistical mechanical formulation of entropy by Boltzmann
1871	Helmholtz joins University of Berlin
1865	Quantitative formulation of Second Law by Clausius
1850	Clausius realizes that there must be two fundamental laws
1847	Clear formulation of First Law by Helmholtz (conservation of energy)
1827	Discovery of Brownian movement
1824	Publication of Carnot's book
1810	Foundation University of Berlin
1799	Ice rubbing Davy
1798	Cannon boring Rumford
1765	Steam engine of James Watt
1712	Steam engine of Thomas Newcomen

Rumford (1753 - 1814)
Carnot (1796 - 1832)
Clausius (1822 - 1888)
Joule (1818 - 1889)
Gibbs (1839 - 1903)
Mayer (1814 - 1878)
Poincare (1854 - 1912)
Boltzmann (1844 - 1906)
Gay-Lussac (1778 - 1850)
Kelvin (1824 - 1907)
Helmholtz (1821 - 1894)
Planck (1858 - 1947)
Nernst (1864 - 1941)
Schrodinger (1887 - 1961)

Appendix VII. Can the Human Body Be Considered a Heat Engine?

The human body can deliver lots of work. Consider, for instance, the athlete running a marathon, or the cyclist racing in the Tour de France. We also know that human body temperature is normally 37°C and that usually the environment is cooler, say 20°C. From this we could suggest that there is some resemblance between a heat engine, in which the body is the heat source, and the cooler environment could act as a heat sink. So let's make a few simple calculations to see how closely the body resembles a heat engine. From Chapter 2, we know that the efficiency of a heat engine is determined by the temperatures of the heat source (the body temperature, T_{body} = 310K) and the heat sink (the environmental temperature, T_{sink} = 293K):

$$Efficiency = \frac{T_{Body} - T_{\sin k}}{T_{Body}} = \frac{310 - 293}{310} = 5.5\%$$

Thus, based on this temperature difference, the body would be able to achieve only 5.5% efficiency. Fortunately, scientific studies already have estimated the human body's efficiency [Whitt et al., 1976] in other ways. One study reasons that for an average man to produce 75 Watts of power, he will need to breathe about one liter of oxygen per minute. That liter of O_2 is combusted in body cells to form carbon dioxide (CO_2). It has also been determined that one liter of oxygen generates in this way about 300 Watts of power. Thus, we can conclude that the efficiency of the human "engine" is 75/300 = 25%. What causes the difference between the 5.5% efficiency as calculated above, and the 25% from the combustion determination? The explanation is that the human body *cannot* be considered a heat engine. The work is not generated in the same way as a steam engine, which directly transforms heat into work and lower-temperature waste heat. Instead, the human body is more like a fuel cell, where chemical energy is transformed into work [Whitt et al, 1976]. For this kind of transformation, one obviously cannot use the efficiency formula of a heat engine.

Appendix VIII. Ways to Concentrate Energy: Nuclear Energy, Photovoltaic Cells, and Fuel Cells

In order for energy to be useful to us, it needs to be available in concentrated form. We know that the total amount of energy stays exactly the same in the universe, and we can only unleash heat or work from available energy when one part of a system contains more energy than another. (For example, consider our solar system: the sun contains massive amounts of energy, but the earth does not). Situations of non-uniform energy distributions have a low entropy; in transforming the energy we distribute it more evenly in the system, and as a result the entropy increases.

Some sources of concentrated energy are easily available, such as fossil fuels and wind power. Some very powerful sources, however, are less accessible, such as nuclear and solar energy. However, by clever manipulation we can concentrate these forms of energy, and make their use practical for us. In the next sections, we will describe the extraction of electricity from nuclear, solar, and fuel cell sources.

Nuclear energy

In the first four chapters, we have discussed extensively essential topics such as energy, heat, and work. We have identified many sources of concentrated energy, including solar, wind, hydrodynamic, electrical, chemical, and fossil fuels. One very different source, which became available just before the Second World War, is nuclear energy. Nuclear energy comes in two different flavors: fusion and fission. Around 1930, it was discovered that under certain conditions, uranium atoms could be split into smaller atoms[137]. But before we immerse ourselves in this phenomenon, I would like to mention two essential features not always recognized but absolutely necessary for nuclear energy to work at all.

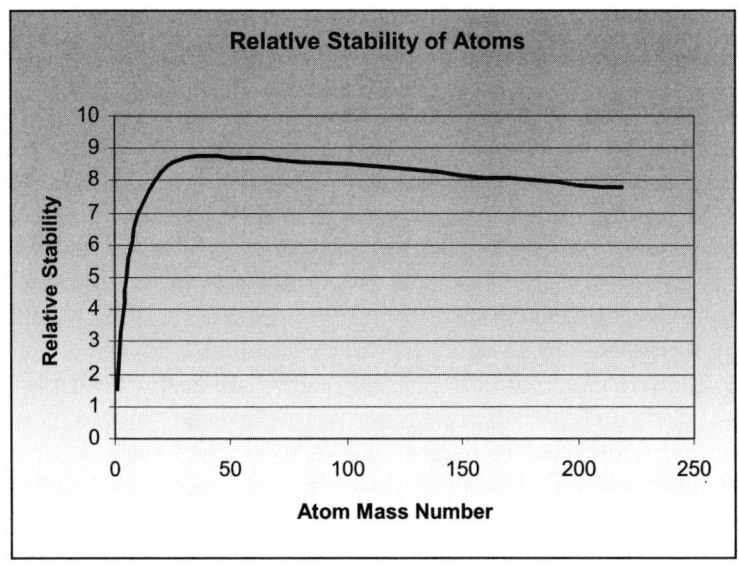

Figure A8.1 Binding energy as a function of mass number (data source NASA).

After Einstein wrote his "Special Theory of Relativity" in 1905 (see Chapter 4), he published another article that same year entitled, *Ist die Trägheit eines Körpers von seinem Energieinhalt abhängig?* (*Does the inertia of a body depend on its energy contents ?*)It was in this paper, reformulated later in 1907, that the famous formula $E = mc^2$ was presented. This formula states that every particle with a mass (m)[138]

---

[137] In nature there are only 92 elements, of which uranium is the heaviest. In an attempt to artificially create more, and thus heavier elements, scientists around 1930 started to bombard uranium atoms with accelerated neutrons in the hope that the uranium nuclei would capture one or more neutrons. This indeed happened, but the resulting atoms became so unstable that they split into smaller atoms. That observation led to the discovery of fission as a source of energy for a controlled nuclear reactor, or eventually, a nuclear bomb.

[138] The mass m is the mass of the particle at rest. From the Theory of Relativity

represents an amount of energy expressed as mc^2. And how much energy would that be? Well, if you took one gram of material[139] and transformed 100% of it into heat, you would get about 9.10^{10} kJ of heat. That's the same amount as would be generated by 3000 tons (3.10^6 kg) of coal! To put it another way, a typical coal-fired power plant can generate about 500,000 kilowatts (kW), or 2.10^9 kJ, per hour. The efficiency of a modern coal-fired power plant can reach nearly 50%, so we need 4.10^9 kJ in heat per hour to generate 500,000 kW. Transforming a gram of material into heat could operate this plant for almost 24 hours!

The realization that mass was equivalent to huge amounts of energy was the first step in discovering the energetic aspects of fission and fusion. However, in order to get access to this source of available energy, we need to understand a second essential feature that is much less well known. That is the phenomenon of the binding energy of protons and neutrons in the atom's nucleus. The binding energy per proton or neutron is determined by, once again, using Einstein's formula, $E = mc^2$. Let's consider an example: A helium atom has two protons and two neutrons in its nucleus. When we add the individual weight of two protons and two neutrons, we arrive at a total mass of 4.03310 atomic units (au[140]). However, experimental work has shown that the weight of a helium atom is only 4.00278 au. The difference of 0.03032 au seems to be small but, when using Einstein's formula, equates to $0.3 \, 10^{10}$ kJ/mole (one mole is about 4 g of helium) of binding energy. When this same calculation is done for all the elements in the periodic system, we come to a remarkable observation (see Figure A8.1). Up to about iron (mass number 57) we notice that heavier elements are more stable than lighter ones. This is the key reason for the fusion process in our sun, where four hydrogen atoms in a thermonuclear reaction fuse to become a helium atom, while losing some mass in the form of radiated energy. But (and get ready to be amazed) for elements heavier than iron, we see the opposite: lighter elements (with atomic mass of 57 or less) are more stable than the heavier elements! This is why nuclear energy works: an

---

we know that the mass of a particle will increase as its velocity increases.

[139] The type of material doesn't matter – it can be water or wood or whatever. This is because at the nuclear (atomic) level, all material is built from the same building blocks: protons, neutrons, and electrons. We will see that it is at this microscopic level where nuclear reactions take place.

[140] One atomic unit, au, is defined as 1/12 of the mass of a carbon atom which is $1.66 \, 10^{-27}$ kg.

unstable uranium atom will fission to yield two lighter atoms[141] and several neutrons[142]: producing about 3 x 10^{-11} J of heat. Per gram of split uranium, this amounts to about 9 x 10^7 kJ of heat!

From these numbers, it's clear why nuclear energy has been so attractive as a source for generating electricity[143]. Its main disadvantage, of course, is the long-term safe disposal of nuclear waste, along with controlling the nuclear reactions (which failed spectacularly at Three Mile Island and Chernobyl). This is especially daunting when you realize that nuclear waste and accidental nuclear discharges can remain radioactive for millions of years!

Now, back to thermodynamics. Despite its novelty, nuclear power is basically another way to convert energy into heat, just as the classical sources have always done. True, the process of obtaining a controlled nuclear reaction is complicated, exacting, and delicate. But assuming that a stable reaction has been created in a reactor, we'll get predictable amounts of heat. From that point forward, the same thermodynamic laws that apply to steam engines also rule nuclear-produced heat. The First and Second Laws will behave just the same as they do in the more classical energy transformations

Under pressure of the Kyoto environmental agreements that call for substantial reductions of greenhouse gases such as carbon dioxide, nuclear energy usage starts to gain in popularity again after years of pressure by environmental movements to abandon this source that liberates available energy (makes energy available at high temperature).

Can photovoltaic cells provide the earth with a sustainable energy source?

In Chapter 5, we briefly touched on the tremendous amount of solar energy that is poured onto our planet and how badly our society

---

[141] The exact product mix depends but is either barium (mass = 141 au) and krypton (mass = 88 au) or strontium (mass = 94) and xenon (mass = 140 au)

[142] Again, the mass of the fragments after fission is slightly smaller than that of the starting uranium atom and neutrons; the difference in mass is converted into heat

[143] Sometimes it looks like we are dealing with a modern kind of perpetual reactor. For instance the *Japan Times* of June 11, 2005 talks about ".....a fast breeder reactor that can generate electricity while producing more fuel than it consumes.........". This sounds great isn't it? It is like a car that produces more gas than its burns while bringing us from point A to point B.

needs an abundant source of clean energy. Most of us have learned the power of solar radiation from a magnifying glass: focus the sun's beams on a piece of paper and you easily burn a hole in it, or even set it on fire! The fact is, we need only convert 0.2% of this energy source into available energy, and the world's energy shortage and many environmental problems[144] will be solved! Of course, we can convert the solar light directly into heat and, with cleverly designed exchanger systems, heat water for domestic usage. Another more practical way would be to convert sunlight directly into electricity. This is now possible by using solar cells, also known as photovoltaic cells. Below, we will describe a few important aspects of this elegant energy conversion technology.

How do solar cells work?

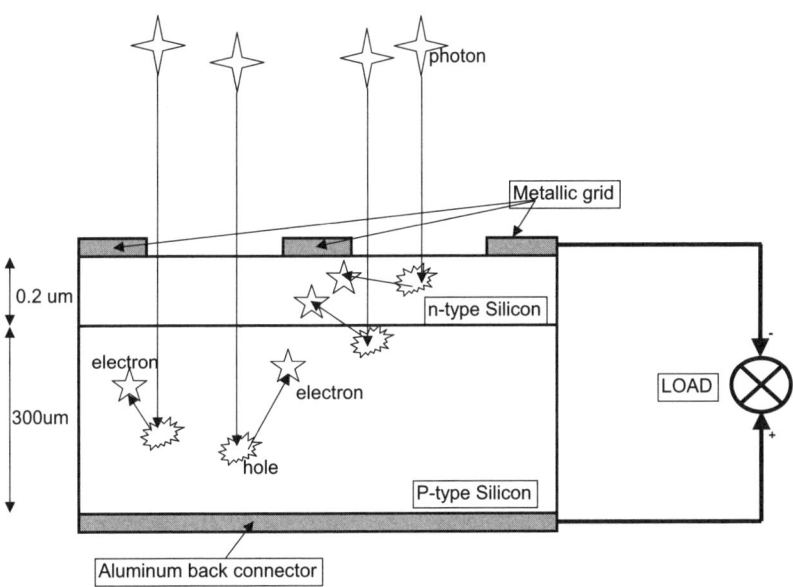

Figure A8.2 (Very) schematic principle of a solar cell.

A solar cell's working principle is based on the properties of semiconductors. While we won't go into details here about the complex

---

[144] See for more information: http://rredc.nrel.gov/tidbits.html

physics of semiconductors, the mechanics of one type of solar cell[145] can be easily understood by looking at Figure A8.2. The starting material is a thin flat piece of silicon about 300 microns thick. By building in a small amount of an impurity like boron, the material can be made into what's called "p-type" silicon. An electrical current in p-type silicon is conducted by positive electrical charges. By exposing the other side of the silicon to phosphorous, a thin layer of about 0.2 micron is converted into n-type silicon, which conducts negative electrical charges. The region where the p-type and n-type silicon meet is called the p-n junction. Now, the p-type side of the silicon is electrically connected to a film of aluminum to form the rear side of the cell. On the front side, above the n-type silicon, a conducting grid of silver-bearing material is laid down to allow the passage of light while electrically connecting to the n-type silicon. The entire assembly is then packaged to resist the environmental elements.

The sun can essentially be seen as a black body that radiates photons, as we have seen in Chapter 4. In this case, the temperature of the black body is somewhere between 5000K and 6000K. At that temperature, the peak wavelength of light is about 0.5 µm ($0.5.10^{-6}$m). What essentially happens in the photovoltaic cell is that the energy of a photon is given to an electron, and that electron goes to the next energy level in the silicon crystal[146]. The wavelength required in order to make that jump possible is smaller than 1.15 µm (remember that the energy of the photon is inversely proportional to the wavelength of the light, see Chapter 4 and Figure A8.3) We are fortunate that many of the photons in sunlight have energies greater than the energy required to enable the electron jump. Once the electrons are at a higher energy level, the p-n junction will push the electrons from the p-type silicon into the n-type silicon, and from there they enter the conductive grid pattern and wire, and then to a useful application. This can be a light bulb (a bit ironic to create artificial light out of sunlight), or a DC motor, or a battery charger to store the power for later use. A solar cell generates electrical power as DC current at about 0.6 Volt. Often, several cells are put in series to raise the voltage to 15V DC. A cell with an active surface area of 10 by 10 cm can generate about 0.2 Watt of electrical power.

---

[145] There are various ways and materials that can be used to build solar cells. Here we consider the most simple one, based a p-n junction in mono- crystalline silicon.

[146] The electron goes from the valance band to the conduction band.

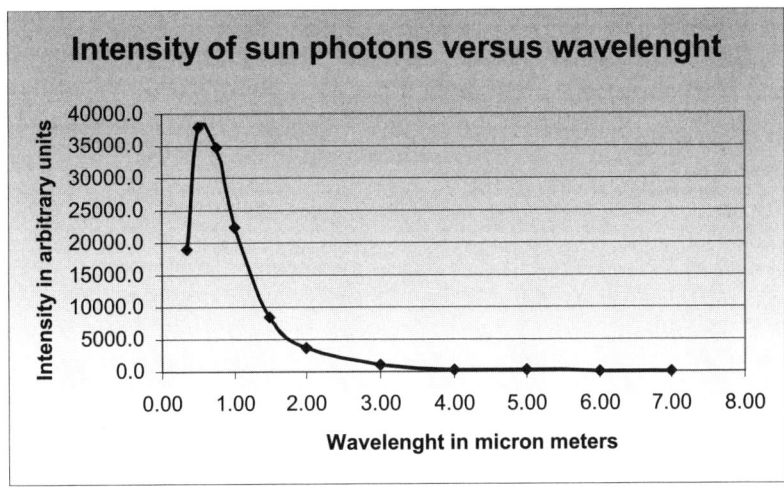

Figure A8.3 Fortunately there is a reasonable match between the energy of the sunlight photons and the absorption wavelength of the silicon (1.2µm)

The principle of the voltaic cell was discovered in 1838 by Becquerel. He observed that the voltage between two electrodes in an electrolytic solution was dependent on the presence of light. In fact, this effect closely resembled the photoelectric effect described by Einstein in his 1905 paper (see Chapter 4). The need for backup power in satellites accelerated the development of solar cells in the 1950s and 1960s. The NASA Vanguard satellite was the first to use solar cells in 1958. Then in 1963, Sharp launched the first commercial solar cells. Ten years later, the oil crisis started a rush toward renewable energy sources, resulting in a demand for solar cells that has increased about 30% in the last few years. In 2002, there were about 2000 megawatts of total solar capacity installed, but this is still less than 0.1% of total worldwide electricity generation. One reason is that the price of solar electricity per kilowatt hour (kWh) is still several factors higher than that generated by fossil fuel or nuclear power plants.

Nevertheless, there are some very strong advantages to using solar power, including:
- Solar power is environmentally benign. Converting sunlight into electrical energy has no impact whatsoever on the environment, and produces no greenhouse gases nor other waste. It has been

argued that the fabrication of solar cells impacts the environment, since hazardous chemicals are used manufacturing them. True enough, but it is also fair to say that the microelectronic and solar industries have been very active in applying containment and abatement techniques to reduce the environmental impact dramatically. As a result, the solar industry has one of the cleanest track records around.
- Solar cells require no maintenance, create no noise, are extremely safe, and last up to 30 years.
- Because they have no moving parts, photovoltaic cells are very reliable.
- Solar cells are truly modular, allowing capacity to be expanded easily when needed.
- Photovoltaic cells are very suitable for developing countries where no power grids exist.

But solar cells also have a few disadvantages:
- Power generation stops when there is no sunlight. This means storage capacity, typically in the form of batteries, needs to be incorporated in a photovoltaic system. Also, since solar cells deliver power as direct current, an inverter is needed to convert DC into AC. The batteries, inverter, and other required components raise the price and lower the reliability of the solar installation.
- As mentioned above, the price of solar per kWh is still rather expensive. However, that must be qualified by the fact that indirect costs are usually not incorporated into the price of conventional power sources. For instance, if the cost of air pollution were counted in the price of electricity from a coal-fired plant, the comparison with solar might be very different. Also, with fossil fuel prices increasing and expected to go even higher because of demand from industrializing nations such as China and India, the price comparison ultimately may favor solar cells. Another factor is that with increasing use of solar power, the economy of scale will lower its cost per kWh. In addition, continuous research and development will further increase the affordability of solar energy.

Appendix VIII

On balance, solar seems like a good idea, and it's fair to say that it should be seriously considered as a viable alternative to established power generating plants. However, one extremely important issue remains: in order to be a viable energy source, the solar cell must produce a *net* amount of energy during its lifetime. Some circles have questioned this, saying that solar cells will always consume more energy that they produce. And indeed, at first sight solar cells look a bit like the perpetual engines that we discussed in Chapter 2. Also, critics point out that fabricating a solar cell and its components consumes power and transforms available energy into the unavailable state. Supporting them are recent detailed studies[147] [Knapp et al., 2000], showing that it can take anywhere between one and five years before a net amount of available energy is generated (this period is called energy payback time, or EPBT). The payback time depends on solar cell design, types of materials used, extent of the manufacturing process, and the climate in which the solar cell is located. But keep this in mind: once the payback is realized, the subsequent solar energy is free! All signs are that we appear to have a truly sustainable energy source that is renewable and environmentally benign.

Fuel cells

We all know that if we burn carbon or hydrogen, heat is produced. In a coal-fired power plant, coal is burned in the furnace, and the resulting heat is used to produce high-pressure steam to drive a turbine that in turn drives an electrical generator. The efficiency of such a power plant is about 30%, with steam formation the limiting factor in the process since we obtain a state of disorder because of the associated thermal, random agitation of the water molecules. And we know what that means: the entropy increases. As we have seen in Chapter 2, the associated losses in efficiency are determined by the factor $T\Delta S$ since the amount of work (ΔW) that can be made available by transforming an amount of energy ΔU is: $\Delta W = \Delta U - T\Delta S$. Thus, the higher the entropy increase, the less transformed energy will be available to generate work.

---

[147] For an overview, see the National Renewable Energy Laboratory website at www.nrel.gov.

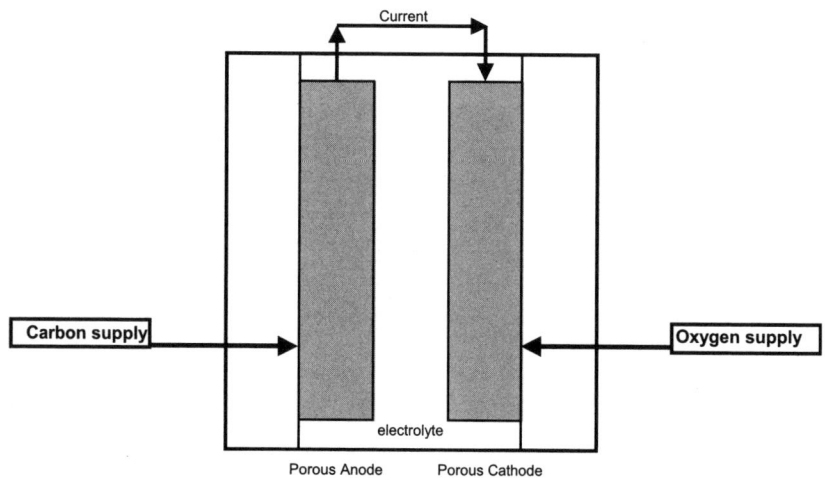

Figure A8.4 Schematic of a fuel cell that burns carbon. Carbon and oxygen are input fuels and CO_2 and electrical power are the output of this cell.

However, if we could eliminate the steam formation step and directly transform the combustion-produced chemical energy into electricity, we potentially could produce much less entropy and increase efficiency. Well, this is indeed possible and the device that can do it is a *fuel cell*. A fuel cell consists of two electrodes immersed in a conducting solution (see Figure A8.4). At the surface of one electrode (the anode, or negative electrode), oxidation reactions occur to produce free electrons, and at the surface of the other electrode (the cathode, or positive electrode), reduction takes place to bind the free electrons. Burning coal is completely unnecessary to this process, whose reactions can be noted as:

Anode: $C + 2O^{2-} \rightarrow CO_2 + 4e$

Cathode: $O_2 + 4e \rightarrow 2O^{2-}$

Net reaction: $C + O_2 \rightarrow CO_2$

Appendix VIII 193

The electrons then flow from the anode to the cathode to drive an electrical device, such as a lamp or a motor. Simple gases, such as hydrogen or methane, are used in place of coal to supply fuel for the reaction because it is not so easy to supply coal to a cell in a simple way. The principle of operation, however, remains the same. For hydrogen the reactions are (if the electrolyte is an acid):

Anode: $2H_2 \rightarrow 4H^+ + 4e^-$

Cathode: $O_2 + 4H^+ + 4e^- \rightarrow 2H_2O$

Net reaction: $2H_2 + O_2 \rightarrow 2H_2O$

It can be shown that at constant temperature and pressure (as is mostly the case in electrical cells), the efficiency, defined as the ratio of electrical energy produced (ΔE) in the cell and that of the heat of coal combustion or the formation of water (ΔH), is equal to:

$$\Delta E/\Delta H = 1 - T\Delta S/\Delta H$$

Herein is ΔH, the heat of combustion of the coal: $C + O_2 \rightarrow CO_2$ or of the hydrogen: $2H_2 + O_2 \rightarrow 2H_2O$, and because heat is produced, ΔH is negative (since heat flows out of the system). If coal is used as fuel for the cell, ΔS is positive and thus the efficiency $\Delta E/\Delta H$ can become greater than 1 — resulting in an efficiency of more than 100%![148] This "free energy" is one of the attractive features of fuel cells.

---

[148] An explanation how that is possible can be found in [Fast, 1962]. Of course we will not be able to generate energy out of nothing. Again, we have here the problem of proper system boundary defintion. If we look only to the fuel cell the efficiency can be higher than 100% (because heat is also taken from the environment), but if we define the system including the surroundings of the cell we are OK and efficiencies smaller (of course!) than 100% will be obtained.

Appendix IX. Qualitative Definitions and Descriptions of Entropy

Because entropy has received so much attention over the years since its conception by Clausius, and has been used in fields other than science, it is not surprising that many different (qualitative) descriptions can be drawn from literature. Below, I have put together a few of these descriptions and their sources.

In sciences:

"Heat always shows a tendency to equalize temperature differences and therefore to pass from hotter to colder bodies."
[early formulation by Clausius, 1850; translation by Max Jammer, 2003]

"It is impossible to extract heat from a reservoir and convert it wholly into work without causing other changes in the universe."
[Formulation by Lord Kelvin]

"It is impossible to construct a perpetual-motion engine of the second kind."
[Formulation by Ostwald, taken from Epstein, 1945]

"The property that describes the randomness or uncertainty is called entropy."
[Howell and Buckius p253]

"Heat cannot be made to go from a low temperature to a high temperature without doing work."
[Stephen Berry, 1991]

"Natural processes are accompanied by an increase in the entropy of the universe."
[Atkins, 1984]

"The key concept of the Second Law is availability, or available energy defined as: availability = the maximum useful work a system can do when it interacts only with its surroundings"

(This description is used to define the efficiency of electrical power production)
[Mitchel, 1983]

"Living systems make every effort for an increase in differentiation whereas non-living systems tend to equalize"
[van Melsen, 1963, my translation]

In non-scientific literature:

"The entropy principle defines order simply as an improbable arrangement of elements, regardless of whether the macro-shape of this arrangement is beautifully structured or most arbitrarily deformed."
[Arnheim, Rudolf, 1971]

"Entropy is a measure of the amount of energy no longer capable of conversion into work."
[Rifkin, 1989]

"A quantity which determines a system's capacity to evolve irreversibly in time, for instance, which describes the propensity of ice lollies to melt and Harry Potter to age. Loosely speaking, we may also think of entropy as measuring the degree of randomness or disorder in a system. When energy is irretrievably converted into heat, then entropy increases (for instance, when Harry falls off his broom with a thud)".
[Highfield, 2002]

Appendix X. Some Simple Calculations and Interesting Numbers

The velocity of gas molecules

The average thermal energy of a gas molecule is 3/2 kT. The average kinetic energy of a gas molecule is $½mv^2$.

From this it can be calculated that the average velocity at room temperature of helium atoms is 1360 m/s, nitrogen molecules is 450 m/s and of carbon dioxide (CO_2) is 411 m/s.

The number of collisions of oxygen molecules at room temperature and atmospheric pressure is about 6×10^9 per second.

Energy

To raise one kilogram one meter requires 10 J of energy.

One pulse of a human heart requires about 1 J.

Energy transformed during one hour of sleep: 150 kJ

One gram of body fat is can yield about 30 kJ

One gram of table sugar yield about 17 kJ

One calorie is 4.184 J

One Btu (British thermal unit) is 1055 J

One calorie dissipated in one cubic centimeter will increase the temperature of the water one degree Celsius.

Burning one kilogram of coal will give approximately 30,000 kJ of heat.

One gram of split uranium gives about $9 \; 10^7$ k J of heat.

Energy of Light in J, eV and wavelength

The wavelength and energy of a photon are related to each other through the formula $E = ch/\lambda$.

Speed of light, c, is 3×10^8 m/s, h is Plank's constant, $6.626 \; 10^{-34}$ J.

One eV is $1.6 \; 10^{-19}$ J, or one MeV is $1.6 \; 10^{-13}$ J

So a photon with an energy of 1 eV has a wavelength of 1.24 microns.

References

Arnheim, Rudolf, *Entropy and Art, an essay on disorder and order*, University of California Press, Berkely and Los Angeles (1971)
Atkins, Peter, *Physical Chemistry* 2nd ed., Oxford University Press, London (1983)
Atkins, Peter, *The Second Law*, Scientific American Library, W.H. Freeman and Company, New York (1984)
Bennet, Charles H. and Landauer, Rolf, "The Fundamental Physical Limits of Computation", Scientific American, p38-46, July (1985)
Bennett, Charles H., "Demons, engines and the Second Law", Scientific American, Nov, p108 (1987)
Berry, R. Stephen, *Understanding Energy, Energy, Entropy and Thermodynamics for Everyman*, World Scientific Publishing Co, Singapore (1991)
Boeker, Egbert, *Een weg naar Utopia?*, Koninklijke Van Gorkum & Comp., BV, Assen (1975), ISBN: 90 232 1306 8
Bushev, Michael, "A Note on Einstein's Annus Mirabilis", *Annales de la Fondation Louis de Broglie*, Vol 25, no 3 (2000)
Burke, James, *Connections*, Little, Brown and Company, Boston (1978)
Bijvoet, Marga J.M., *Art as Inquiry*, Peter Lang Publisher Inc., (1997)

Clausius, Rudolf, "Über die bewegende Kraft der Wärme und die Gesetze welche sich daraus für die Wärmelehre selbst ableiten lassen", *Poggendorffs Annalen der Physik und Chemie,* **79**, 368-397 and 500-524 (1850)

Collier, John, "Two faces of Maxwell's demon reveal the nature of irreversibility", *Studies in History and Philosophy of Science,* Vol 21, no 2, p257 (1990)

Davis, Paul in *Entropy and Information in the Physical Sciences,* p 11; L Kubát and J. Zeman editors; Elsevier, Amsterdam (1975)

Dulken van, Stephen, *Inventing the 19th Century. 100 inventions that shaped the world,* The British Library Board, ISBN 9076988218 (2000)

Einstein, A., *Annalen der Physik,* **11**, 170 (1903)

Ebeling, W. and Hoffman, D, " Grand Schools of Physics The Berlin School of Thermodynamics founded by Helmholtz and Clausius", *European. J. Phys.,* 12,1-9 (1991).

Epstein, Paul S., *Textbook of Thermodynamics,* John Wiley & Sons, Inc., New York, 3rd printing (1945)

Etheridge, D.M., L.P. Steele, R.L. Langenfelds, R.J. Francey, J.-M Barnola, V.I. Morgan; "Natural and anthropogenic changes in atmospheric CO2 over the last 1000 years from air in Antarctic ice and firn" *J. Geophys. Res.* 101 (D2) 4115-4128 (1996)

Etheridge, D.M., L.P. Steele, R.L. Langenfelds, R.J. Francey, J.-M. Barnola and V.I. Morgan; "Historical CO2 records from the Law Dome DE08, DE08-2 and DSS ice cores.' In *Trends: A Compendium of Data on Global Change.* Carbon Dioxide Information Analysis Center, Oak Ridge National Laboratory, Oak Ridge, Tenn, USA. (1998) see also: http://www.cdmc.esd.ornl.gov/trends/co2/lawdome.html

Faber, Malte; Niemes, Horst and Stephan, Gunter; *Entropie, Umweltschutz und Rohstoffverbrauch*; Spinger Verlag, Berlin, (1983)

Fast, J.D., *Entropy,* McGraw-Hill Book Company, Inc., New York (1962)

Flamm, Dieter; "Ludwig Boltzmann – A Pioneer of Modern Physics"; XXth International Congress of History of Science, July 25 (1997) Liège, Belgium

Flamm, Dieter, "Einführung zu Ludwig Boltzmanns Entropy und Wahrscheinlichkeit", this is an introduction of *Entropie und Warscheinlichkeit, 1872-1905* von Ludwig Boltzman in

Ostwalds Klassiker der Exakten Wissenschaften, Band 286, Verlag Harri Deutsch, Frankfurt am Main (2000). This book is contains a nice compilation of the most important articles from Boltzmann in original version.

Frank, Michael P., "Physical Limits of Computing", *Computing in Science & Engineering*, Vol 4, no 2 (2002)

Galison, Peter, *Einstein's Clocks, Poincaré's Maps, Empires of Time*; W.W. Norton & Company, Inc., New York (2004)

Gallager, Robert G., "Claude E. Shannon: a Retrospective on His Life, Work, and Impact", *IEEE Transactions on Information Theory*, Vol. 46, no 7, p. 2681 (2001)

Glencoe, *Biology the Dynamics of Life*, McGraw-Hill (2004)

Georgescu-Roegen, Nicolas, *The Entropy Law and the Economic Process*, Harvard University Press, Cambridge, Massachusetts (1971)

Hasenöhrl, F., ed. *Wissenschaftliche Abhandlungen von Ludwig Boltzmann*, Vol I-III, J.A. Barth, Leipzich (1909)

Helmholtz von, Hermann; "On the Conservation of Force", Carlsruhe, (1862)

Highfield, Roger, *The Science of Harry Potter*, Penguin Books, London (2002)

Howell, John R., Buckius Richard O., Fundamentals of Engineering Thermodynamics, McGraw-Hill, Inc., New York (1992)

Jammer, Max, "Dictionary of the History of Ideas", www.etext.lib.virginia.edu, University of Virginia Library (2003)

Kleidon, A., Lorenz, RD, *Non-equilibrium Thermodynamics and the Production of Entropy: Life, \Earth and Beyond (Understanding Complex Systems)*, Springer Verlag, Heidelberg 92004)

Klein, Martin J., "Thermodynamics in Einstein's Thought"; *Science*, Vol 157, 509 (1967)

Knapp, Karl and Jester, Theresa; "An Empirical Perspective on the Energy Payback Time for Photovoltaic Modules"; Solar 2000 Conference. Madison, Wisconsin, June 16-21 (2000)

Krishnan, Rajaram; Harris, Jonathan and Goodwin, Neva in "A Survey of Ecological Economics; Island Press, Washington D.C., p186 and p 191 (1995)

Lambert, Frank L., "Shuffled Cards, Messy Desks, and Disorderly Dorm Rooms – Examples of Entropy Increase? Nonsense!, *J. Chem. Educ.*, 76, 1385 (1999)

Lebowitz, Joel, *Physics Today*, September, p32, 1993

Lorenz, Ralph et. al., "Titan, Mars and Earth: Entropy Production by Latitudinal Heat Transport", *Geophysical Research Letters*, Vol. 28, No 3, p. 415-418 (2001)

Manning, Richard, "The oil we eat: following the food chain back to Iraq", *Harper's Magazine*, February 2004

Melsen van, A.G.M., *Natuurwetenschap en Techniek, een wijsgerige bezinning*, Het Spectrum N.V., Utrecht (1963)

Mendoza, E., "A Sketch for a History of Early Thermodynamics", *Physics Today*, Sept., p32 (1961)

Mirowski, Philip; *Against Mechanism, Protecting Economics from Science*, Rowman & Littlefield Publishers, Lanham, MD (1988)

Mitchel, John W., *Energy Engineering*, John Wiley & Sons, New York (1983)

Moore, Walter, J., *Physical Chemistry*, Longman, London, (1978)

Novem, *Wereld in wording*, Van Goor zonen, Den Haag (1972)

Rees, William E., "Revisiting Carrying Capacity: Area-Based Indicators of Sustainability", *Population and Environment: A Journal of Interdisciplinary Studies*, Volume 17, no 3 (1996)

Rifkin, Jeremy, *Entropy into the greenhouse world*, Bantam Books, New York, (1989)

Riordan, Michael and Hoddeson, Lillian, *Crystal Fire, the birth of the information age*, W.W. Norton & Company, New York (1997)

Schlegel, Richard, "Entropy and Relativity Theory" in *Entropy and Information in Science and Philosophy*, Kubat and Zeman ed., Elsevier Scientific Pub. Co., Amsterdam (1975)

Schrödinger, Erwin, *Science Theory and Man*, New York: Dover Publications (1957)

Schrödinger, Erwin, *What is life?*, Cambridge University Press, London, (1951)

Shannon, C.E., "A mathematical theory of communication (Part 1)", *Bell System Technical Journal*, vol 27, pp. 379-423 (1948); Part 2 appeared in the same volume on pp. 623-656.

Sharma. A. Abstract International conference Physics 2005, A century after Einstein (Institute of Physics, London Bristol), p144 (2005)

Walker, Davis, *Energy, Plants and Man*, Oxy Graphics, England, University Books, Mill Valley, CA (1993)

Wang, G.M. et. al., " Experimental Demonstration of Violations of the Second Law of Thermodynamics for Small Systems and Short Time Scales", *Physical Review Letters*, vol 89 (3), (2002)

Wheeler, Lynde Phelps; "Josiah Willard Gibbs, *The History of a Great Mind*, Yale University Press, New Haven (1952)

Whitmarsh J. and Govindjee; *Encyclopedia of Applied Physics*, Vol 13 (1995)

Whitt, F.R. and Wilson, D.G., *Bicycling Science*, MIT Press, Cambridge (1976)

Index

absolute temperature, 20, 22, 34, 66, 68, 173
adenosine, 115, 117
adiabatic, 28, 30, 38, 40, 68
air pollution, 98, 105, 190
air pump, 38, 167
Aristotle, 5, 9, 52
AT&T, 138, 140, 141
atomos, 52
ATP, 133, 134
Augustine, Saint, 88
Avogadro, 53
Ayres, 95
Bardeen, 140
Becquerel, 71, 189
Bell, 138, 140, 141, 202
Bell Telephone Laboratories, 140
Bhaskara, 45, 46
Bible, 154, 155
Big Bang, 136
binary digits, 145
black body radiation, 73
Boltzmann, 52, 54, 59, 60, 64, 65, 66, 68, 70, 71, 72, 81, 82, 83, 84, 87, 91, 92, 104, 116, 137, 148, 149, 150, 158, 161, 162, 173, 200, 201
boring of cannon barrels, 24
Brattain, 140
breeder reactor, 186
Brillouin, 152
Brogly, 86
Brown, 61, 64, 199
Brownian movement, 61, 62, 83, 114, 117, 153
Calvin cycle, 133
Cape Cod, 139
carbon dioxide, 98, 130, 131, 132, 134, 135, 181, 197
Carnot, 13, 28, 29, 31, 32, 33, 34, 35, 37, 38, 39, 40, 102, 161, 162
Carnot cycle, 29, 32, 34
carrying capacity, 100
Celsius, 14, 21
CFCs, 98, 99
channel, 140, 142
chaos, 1, 5, 6, 54, 56, 57, 62, 114, 171
Chernobyl, 100, 186
chlorofluorocarbons, 98
chlorophyll, 132

Christian religion, 154
chromosomes, 115
Clausius, Rudolf, 6, 14, 15, 28, 34, 36, 39, 72, 165
Club of Rome, 95, 99, 100
combustion, 15, 29, 99, 132, 134, 181
Compton, 86
concentration gradient, 117, 173
condensor, 11, 34
conservation of money, 102
convection, 18, 127
cooling machine, 34
Copernicus, Nicolaus, 9
cytosine, 115, 117
d'Alembert, 23
da Vinci, 45
Dalton, 53
data compression, 143, 144, 145, 147, 148, 150, 159
decay, 120, 154, 159
Democritus, 52
determinism, 157
Diesel, 168
diffusion, 110, 117, 127, 128, 129, 173, 174
DNA, 5, 113, 115, 117, 155
dynamos, 5
economic process, 13, 95, 96, 101, 102, 103, 106, 110
efficiency, 5, 12, 13, 16, 29, 32, 33, 36, 37, 39, 40, 48, 49, 69, 70, 101, 102, 106, 107, 109, 111, 116, 130, 131, 153, 161, 162, 181, 185
Einstein, 14, 54, 62, 68, 71, 73, 78, 80, 81, 82, 83, 84, 85, 86, 87, 89, 91, 150, 162, 184, 185, 189, 200, 202
Elvius, 21
Energy, 3, 4, 5, 6, 7, 13, 14, 15, 20, 23, 24, 25, 26, 27, 28, 29, 30, 32, 34, 35, 38, 40, 43, 44, 45, 47, 48, 49, 51, 62, 65, 68, 69, 70, 71, 72, 73, 75, 76, 77, 78, 81, 82, 83, 84, 85, 87, 91, 96, 99, 100, 102, 104, 105, 106, 108, 109, 110, 113, 114, 118, 121, 122, 123, 124, 125, 126, 127, 129, 130, 131, 132, 133, 134, 135, 136, 149, 153, 154, 155, 156, 157, 158, 161, 162, 165, 171, 173, 175, 176, 181, 183, 184, 185, 186, 187, 188, 189, 191, 196, 197, 198
entropy production, 42, 43, 47, 98, 107, 108, 111, 127, 128, 134
Epstein, 13, 15, 195
equilibrium, 17, 19, 20, 22, 41, 114, 135, 136
ether, 73, 78, 79
Evolution, 155
extensive parameters, 165
Fahrenheit, 21
Faraday, 53, 84
Ferdinand II, 20
First Law, 13, 25, 27, 35, 43, 44, 47, 70, 91, 154, 158, 161
fixation, 130, 132
flux, 124
food chain, 121, 129, 130, 131, 135, 202
French Cable Station, 139
French Cable Station Museum, 139
friction, 15, 38, 40, 49
Galilei, Galileo, 9
gas mileage, 4
Gay-Lussac, 21, 22
genes, 116, 117
Georgescu-Roegen, 6, 95, 101, 106, 109, 201
Gibbs, 165, 203
global warming, 98
glucose, 131, 132, 133, 134
God, 154, 158
Greenhouse, 95, 98
Greenwich time, 88
guanine, 115
Hafele and Keating experiment, 90
heat capacity, 163
Heat Death, 42, 113, 135, 152, 158
heat motion, 115
heat transfer, 25
Helmholtz, 14, 15, 72, 135, 200, 201
Hertz, 77, 138
human body, 115, 117, 120, 181
hydrochloric acid, 53
Industrial Revolution, 6, 10, 65, 99
intensive parameters, 165
Internet, 138, 155
iron ores, 110

Index

irreversible, 15, 16, 18, 19, 36, 37, 39, 40, 41, 43, 66, 70, 95, 98, 102, 111, 155, 173, 175, 176
isolated system, 3, 41, 42, 48, 85, 102, 104, 105, 107, 108, 113, 118, 121, 155, 156, 157, 173, 175, 176, 177
isothermal, 28, 30, 41, 63
Japan Times, 186
Jeans, 76
Joule, 14, 26, 40, 42, 59, 131, 134, 162, 197
Kelvin, 21, 35, 59, 68, 73, 121, 195
Kepler, Johannes, 9
King, Alexander, 99
Kirchhoff, 83
Koopmans, 95
Kyoto's environmental agreements, 186
Le Direct, 139
Lebowitz, 68, 202
Lenard, 77, 78
Los Alamos Scientific Laboratory, 149
Loschmidt, 66
lossless compression, 146
Mache, 68
macrostates, 57, 58
macrosystem, 65, 67, 149
Malthus, 95
Marcon, 140
Marconi Station, 140
Maximum Entropy Principle, 128
Maxwell, 51, 84, 85, 138, 149, 150, 151, 152, 154, 171, 200
Maxwell's demon, 149, 150, 151, 152, 153, 154, 200
Mayer, 14, 15, 132, 162
memory, 143, 144, 153, 154
MEMS, 63, 64, 69
meridional heat diffusion coefficient, 127
methane, 98
Michelson-Morley, 73, 79
microelectromechanical, 63
microstates, 57, 58, 59, 60, 65, 82, 92, 104, 115, 149
modern physics, xvi, 71, 81, 91
Morse, 138
Morse Code, 139
MP3, 146, 148, 150
MPEG, 146, 148
NADPH, 133
NASA Vanguard satellite, 189
Nernst, 14, 67, 68, 72
network theory, 140
Neumann, John von, 149
Newcomen, 11, 29
Newton, Isaac, 5, 10
nitrous oxide, 98
Nobel Prize, 43, 72, 77, 78, 87, 142, 157
noise, 140, 190
nuclear energy, 183, 186
Ohm, 72
origin of heat, 48
Ostwald, 52, 53, 68, 82, 195
Otto engine, 168
ozone layer, 98, 99
Peccei, Aurelio, 99
perpetual motion machine, 35, 44, 85
photoelectric effect, 73, 77, 83, 189
photophosphorylation, 132, 133
Photosynthesis, 132
photovoltaic cells, 121, 183, 186, *187*
Planck, 14, 54, 68, 71, 73, 75, 76, 78, 81, 82, 83, 84, 86, 87, 91
Poincaré, 66
polar regions, 127, 128
postulate, 25, 44, 80, 81
power plant, 185
Prigogine, 157
prions, 115
probability, 54, 57, 60, 71, 81, 82, 87, 92, 142, 149
quanta, 71, 75, 76, 83, 86, 131, 132
quanten, 83
quartz, 104
Rayleigh, 76
recycling, 98, 103, 104, 105, 106, 108, 111
refrigerator, 34, 35, 99
renewable, 96, 98, 109, 189, 191
reversible, 15, 16, 18, 19, 30, 31, 36, 37, 38, 39, 40, 41, 42, 43, 66, 111, 173, 175, 176
Rey, 20
Rifkin, 95, 109, 196, 202
Rontgen, 71
Rumford, 14, 15

Savery, Thomas, 10
Schrödinger, 14, 72, 86, 87, 114, 116, 120, 202
Second kind, 35
Second Law, xvi, 28, 34, 35, 36, 42, 43, 47, 48, 52, 60, 62, 64, 73, 81, 91, 95, 108, 110, 113, 118, 120, 129, 135, 152, 154, 155, 157, 158, 159, 161, 162, 175, 176, 177, 186, 199, 203
Semat, 71
Shannon, 137, 140, 141, 142, 143, 144, 145, 147, 148, 149, 150, 153, 158, 159, 201, 202
Shannon entropy, 141, 143, 145, 148, 149, 150, 159
Sharp, 189
Shockley, 140
silica, 104
Smithson, Robert, 158, 159, 160
Smoluchowski, 152
solar cell, 109, 135, 187, 188, 191
Space Shuttle, 80
spermatozoa, 115
statistical variations, 114
Steam engines, 5, 6, 25, 29, 32, 33, 34, 48, 186
Stefan, 83
Stevin, Simon, 45
Stirling approximation, 149
sunlight, 98, 99, 121, 124, 127, 131, 134, 135, 187, 188, 189, 190
Szilard, 152
telegraph, 65, 88, 138, 139
TELEX, 138

thermal equilibrium, 136, 163
thermometer, 21, 22, 23
thermos, 5
Third Law, 43, 67, 163
Thompson, 14, 24
Thomson, 13, 71
Three Mile Island, 100, 186
thymine, 115
TIFF, 146
time, xv, 4, 6, 7, 10, 12, 13, 14, 18, 20, 21, 24, 25, 33, 34, 40, 45, 47, 52, 62, 64, 65, 66, 68, 70, 72, 73, 78, 80, 82, 83, 84, 85, 86, 88, 89, 90, 96, 97, 107, 108, 110, 138, 139, 140, 141, 142, 161, 191
time arrow, 47, 65
Tolman, 91
transatlantic cables, 142
transmission, 138, 139, 140, 144, 159
trope, 6, 15
ultra-violet catastrophe, 76
universe, 1, 6, 8, 9, 35, 42, 88, 90, 115, 116, 121, 122, 123, 127, 129, 135, 136, 157, 161, 162, 165, 183, 195
uranium, 110, 183, 184, 186, 197
Vancouver, 101
Watt, James, 11, 33
Western Union, 109, 138
Wien, 83
wireless, 138, 139, 140, 142
world population, 95, 96, 100, 101
World Wide Web, 138
Zermelo, 66
Zeroth Law, 43, 163